THE ACCEPTABILITY OF RISKS

The logic and social dynamics of fair decisions and effective controls.

Published by Barry Rose (Publishers) Ltd. in association with the Council for Science and Society

Brown and Son (Ringwood) Ltd., Crowe Arch Lane, Ringwood, Hants.

SBN 85992 107 7

TABLE OF CONTENTS

PREFACE

The risk of suffering physical harm is an inescapable aspect of living. Over recent generations, scientific and technological advances have reduced or eliminated many risks which were previously commonplace. But they have also created new ones. Ideas, too, have changed. Children no longer work in coal mines or as chimney sweeps, and we welcome as liberal and humane the laws which have put a stop to these practices. Yet, paradoxically, all our babies are now exposed to the effects of radioactive strontium in their milk. Some new risks have become more pervasive, and harder to define, as well as to eliminate.

No one can totally eliminate all risks of physical harm, nor is it necessarily praiseworthy for individuals or for governments to impose that ideal as a policy. It would be a dull world in which school-children were forbidden to climb trees or to play with hard balls; in which mountain climbers were no longer allowed to attack the Eiger or Everest, or astronauts to venture into space. The concept of freedom must entail a degree of freedom to risk *one's own* life and limb. But this liberty is not a licence to take risks with the lives and limbs of *others*. It is still less justifiable where those subjected to serious risk have no practical means of self-protection or control. The debate on civil nuclear power now hinges on just such an issue. The harm from pollution by a release of radioactive poisons could be very great; but the probability of such an event is claimed to be very small. Is such a risk a serious one? Is it acceptable, when weighed against the promised benefits? And who should accept it, and for whom?

Any genuine discussion of risk soon encounters questions like these, involving costs, benefits, probabilities and notions of acceptability and, above all, questions of choice. Some elements of the discussion will be based on scientific research. Calculating the frequency of past occurrence and the probability of future recurrence of particular harmful events, for example, often requires an application of the scientific method. Personal preferences are important; we all take risks and in some way justify them against the personal benefits we expect. However, in the last resort, the control of risks is the responsibility of government; people's behaviour must be limited in the interest of others, and this requires sanctions that only the State possesses.

Naturally, a government's ideas of acceptability of risk change; they are moved by the weight of opinions, and the tide of political events. Governments are sometimes slow to recognize new risks as they arise, and are often reluctant to take positive steps to tackle them. This reluctance to alter the criteria of acceptability of risk in the face of economic and social repercussions opens the door to controversy and debate. How should debates on the acceptability of risk be conducted? Can they be reasoned, or will there only be confrontations between incompatible opinions? Can there be a consensus on the problem and on the principles of the discussion, or only power-struggles between various corporate interests and militant pressure groups? This is an important problem for the effectiveness, indeed the survival, of democratic government in a technological age.

With this problem in mind, the Council for Science and Society proposed the topic of 'the standards to be applied, and the procedures for applying them, in weighing "acceptable risk" against anticipated benefits from new technologies'. The Council recognized that there was as yet very little expertise on this topic, since even the scientific assessment of risk has been a neglected discipline. It was hoped that out of an investigation by one of its Working Parties there might emerge a useful study which would not so much describe and analyse all existing practices, but rather provide a framework for the use of those involved in research, in debates, and in decisions on the acceptability of risks.

This report was unanimously approved for publication by The Council for Science and Society at its meeting of September 29, 1976. We recommend it as an original contribution to the important and growing public discussion on the risks associated with technological development.

MICHAEL SWANN

THE WORKING PARTY

The members of the Working Party were:-
Brigadier R. L. Allen, CBE (former Chief Inspector of Land Service Ammunition).
Eric Burhop FRS (Professor of Physics, University College, London. Council Representative)
Dr Mike Flood (Consultant, Friends of the Earth)
Jonathan Glover (Philosopher, Fellow of New College, Oxford)
Jerry Ravetz (Secretary)
Anthony Woolf (Chairman, Lawyers' Ecology Group)

The Working Party also wishes to give particular thanks to the following: Mr. F. R. Farmer of the U.K.A.E.A. Safety and Reliability Directorate; Gordon Atherley, Professor of Safety and Hygiene at the University of Aston; and Mr. Octavius Critchley, then a Senior Research Fellow in the University of Manchester on leave of absence from the Nuclear Installations Inspectorate.

We are grateful for information and advice to the Working Party given in a personal capacity by Mr Philip Wilson-Dickson (HM Inspector of Fire Services, Home Office) and Dr Walter Tye (former Controller Safety, Civil Aviation Authority).

Others we should like to thank for offering their views on the subject, for providing useful information, and for answering our queries, include Dr John Adams; Dr Charlie Clutterbuck, Mr Barry Turner (lecturer in Sociology, University of Exeter); Mr D. V. Warren (Civil Aviation Authority); and Mr Bob Swindale.

We also wish to extend our gratitude to Ms. Margaret Hebblethwaite and Mr Bruce Page for their participation in the early discussions of the Working Party.

Successive drafts of the Report were prepared for discussion and modification by Jerry Ravetz, the Executive Secretary of the Council for Science and Society. In all phases of this work he was given enthusiastic support by Mr Ian Riley (then Assistant Secretary of the Council), Mr John Hillier (of the Council's staff), and the Council's typists.

INTRODUCTION

In approaching the problem of acceptability of risk, we on the Working Party were struck by the great variation in ideas and practices among the different agencies that regulate risks, and also by the small scale and fragmentation of research. The problems of risk are complex; many criteria are not easy to quantify, and are of doubtful relevance; and many of the beliefs prevailing among experts seem subjective. Even the mere task of classificiation of the different procedures for judging the acceptability of risk seemed daunting to us. If these difficulties were encountered by a Working Party of mixed disciplines such as ours, how much harder must it be for inexpert participants in a public discussion to advance their case competently?

As the analysis developed, we worked through various conceptions of our subject, and it may be useful to record these. The initial topic suggested to us was "Acceptable Risk: The standards to be applied, and the procedures for applying them, in weighing acceptable risk against anticipated benefits from new technologies". This seemed to call for a simple listing of criteria on both sides of a balance, and might have been accomplished by a review of several case studies on major areas of risk. But several difficulties blocked this approach. Risks have been largely neglected as a field of social or historical research. Comprehensive, reliable surveys simply do not exist. The reports of official inquiries into particular accidents, excellent in their own way, do not provide the breadth and general relevance that we need. We considered organizing our own research, and indeed carried out some limited studies into particular areas. From this experience we saw that useful full-scale studies suitable for analysis by a working party would require resources of time, funds and management beyond our means. We accordingly decided to proceed more informally, relying on the range of experience represented on the Working Party, which was supplemented by correspondence and discussion with experts in a variety of fields. Several members eventually contributed accounts of the areas where their knowledge is greatest. These comprise the bulk of the Appendices; they are referred to wherever appropriate in the text, and we hope that they are themselves a useful contribution to the literature on risks. For empirical data and theoretical ideas on risks, we derived great benefit from the American study by William W. Lowrance (1), who has made one of the first attempts at a wide-ranging investigation in the uncharted field of risk-acceptance.

In our study, we start by introducing the problem and then making some clarification of our ideas. This is necessary to avoid the confusions that have hampered other studies in this field. We then examine the "scientific" side of the problem, in the assessment of risk. The "human" side, in the evaluation of risks, comes after. At every stage, we are concerned to see what can legitimately be measured or discussed objectively and where are the limits to this approach. We then survey the important differences among the various sorts of "acceptability" of risk. Finally, we analyse what is involved in achieving "fair" decisions on risk and effective controls.

Our analysis tends to be critical but, we hope, constructively so. The general theme that emerges at every stage is that while scientific methods and scientific rigours are essential for proper analysis of risk, the human factor is always present and must be respected and utilized in all analysis, decisions and controls, for risks to be genuinely acceptable to the people concerned.

1. THE PROBLEM OF RISK — SOME EXAMPLES

The problem of risk is now an important component of all decisions about technological development. Yet the concept is surprisingly difficult to grasp. We all incur risks, and we impose them on others in our ordinary lives, often quite unawares. Popular concern about risks has increased recently, as man-made disasters carry the threat of even worse to come from new or projected technologies. The management of risk involves far more than the calculation of probabilities of the occurrence of accidents or disasters. Scientific methods for the assessment of risk need to be still further developed and strengthened; but final judgments on the "acceptability" of risks require a multi-disciplinary approach. We can best illustrate the complexity of the problem by some familiar examples.

1.1 Smoking

Anyone who regularly smokes cigarettes exposes himself to a calculable risk of early illness and death. The harm is not restricted to him and those directly dependent on him; ill-health is a burden on society as a whole. The smoker presumably finds the risks "worthwhile" when weighed against the benefits to him. Society tolerates the risk as "unavoidable" because of the costs of an attempted prohibition on cigarette smoking. But smoking is not regarded as a "negligible" cost on society, and hence various measures are adopted to discourage it. Thus, the risks that people and societies judge to be "acceptable" will depend only partly on scientific evidence. The various benefits of any dangerous practice, and the costs of stopping or changing it, must necessarily be included in the judgment.

1.2 Air Travel

Risks are not simply a given, unalterable part of the natural world; their severity can be, and often is, changed by the intelligent application of the results of scientific studies. Although an individual's chances of suffering a fatal accident are higher while flying than while travelling by land or sea, the overall level of risk in air travel decreases steadily from year to year. Much of the credit for this lies with the regulatory agencies which have developed scientific analyses of the main human and mechanical causes of accident. Armed with comprehensive statistics they have been able to demand and achieve tighter design targets for successive generations of aircraft. Today the risk encountered by passengers on each flight is near to the level apparently considered "negligible" in ordinary actions, and therefore acceptable to the great majority. (See Appendix 1).

1.3 Flixborough

No assessment of risk is complete if it is limited to a consideration of the statistical properties of components of hardware. Management practices, operation procedures and human failings are also prime causes of risks. The explosion of the Nypro chemical plant at Flixborough near Scunthorpe, in 1974, alerted the public to the existence of "major hazards" — the situation where an accident could cause a holocaust and threaten the lives of thousands of people. One way of investigating how such accidents occur (a way perhaps over-emphasized by the Commission of Inquiry into the Flixborough disaster), is to focus on the immediate physical events which triggered the disaster. In the case of Flixborough such an approach led to a concentration on the question of which of two pipes was the first to rupture. Yet, in an important sense, this was only part of the story: indeed a small part. A full explanation of an accident requires more than the identification of the initiating events. It involves an understanding of the effects of such management practices as running the plant without a qualified mechanical engineer on site, and of storing large quantities of hazardous chemicals close to potentially dangerous areas of the plant. (See Appendix 3).

1.4 "Trans-science"

As we become increasingly aware of the complex effects of industrial technologies, new and old, we must expect to venture frequently beyond the limits of the straightforward scientific assessment of risk. The difficulties experienced by those seeking such estimates are well illustrated by the example of chronic exposure to low doses of pollutants, especially where symptoms may take some time to appear. It is theoretically possible to assess the carcinogenic effects of toxic by-products leaking from a factory, even when they are present only in minute quantities. But to perform a statistically significant test of the hypothesis in a laboratory might require enormous numbers of experiments to be carried out — perhaps several thousand million. Such problems have been labelled "trans-scientific" (2): their solution is *beyond* the practical application of the scientific method, given the limits of financial and human resources. In such cases, science can provide at best an indication of the result, not a full specification. Applying mathematics to such contingencies requires great skill and care (see Appendix 7). Society must then, in the final analysis, rely on its judgment in balancing imponderables.

1.5 Asbestos

The control of risks involves more than a scientific knowledge of their causes. An effective corps of "guardians" is also necessary, lest short-term considerations of profit or convenience come to dominate the judgments of those who create risks. The recent disclosures of asbestos-caused cancers among industrial workers illustrate how an official Inspectorate may be well aware of the facts on hazards, and yet unable to act on them. The carcinogenic properties of asbestos dust were known for decades, while Inspectors continued to allow excessive concentrations to exist (see Appendix 5). Ensuring that the "guardians" do their job involves an appreciation of what we shall call the "social dynamics" of a hazard situation.

1.6 Conclusion

From these selected examples, we can see that scientific knowledge of the causes of risk is only a part of the problem of control. Personal tastes, management practices, and the effectiveness of monitoring agencies can all strongly influence the use that is made of scientific findings, even when these are available. We need a conceptual framework in which all the scientific and social aspects of risks can fit. Otherwise our judgments about risks will remain confused, and our controls inadequate. In this report we attempt to sketch such a framework of ideas; and our conclusions will show it in use.

2. HAZARDS AND RISKS, CAUSES AND EFFECTS

It will be useful to make a brief review of standard terms and concepts, so that we may later use them freely in discussing the various problems of the management of risks.

The terms "hazard" and "risk" have acquired a variety of meanings in the literature. To avoid confusion, we will use the terms in the following senses:

"Hazard" will describe a situation with the potential to cause harm (to people, property or the environment).

We will use "risk" to refer to the probability that the potential of a hazard will be realized. Following common usage, however, we will also let "risk" stand for a combination of the probability with the harm itself. This reflects the slight distinction between "hazardous" and "risky", the former referring more to the situation and the latter more to the harm. But we have no single phrase to describe adequately what happens when harm is sustained. In some cases we may refer to the event as an "accident"; but we do not use that term for, say, diseases caused by pollution. Hence we will often use the term "realized hazard" or sometimes "harmful event".

2.1 A Caution on "Safe"

At this point we must also resolve a difficulty caused by the common use of the term "safe".

To say that a situation is "safe" implies a final judgment that the risk is in some sense "acceptable" or even non-existent. But experience with medical and industrial hazards, such as thalidomide, asbestos and radioactive pollution, shows that harm can occur when it is not expected and that a risk once deemed "acceptable" may at any time need to be reconsidered.

Even though the term "acceptable" has many problems of its own (see Chapter 5), it does, at least, refer to the human judgments of the situation, and thereby has the connotations of being tentative and fallible. By contrast, "safe" refers only to the situation itself, and accordingly seems to entail absolute finality.

In his study of risk, Lowrance tried to retain the word "safe" by means of a definition, but in doing so he tended to confuse the issue. His definition is: "A thing is safe if its risks are judged to be acceptable" (3). A more precise definition would be "A thing is *provisionally categorized as* safe if its risks are *deemed known*, *and in the light of that knowledge* judged to be acceptable". This rescues his definition from its obvious paradox: a thing may be "safe" for a user who is ignorant of its hazards, while being considered "dangerous" by another who is aware of them. But then there is no particular benefit in using the word "safe", since it is either reduced to "acceptable" or retains a misleading connotation of finality. We shall therefore avoid the use of "safe" and also be sparing in our use of "safety" except where there is no possible confusion.

2.2 What are "Hazards" and "Risks"?

Risks are as variable as life itself. It is therefore very difficult to classify them into a form suitable for shaping or guiding policy. We can see the scope of the problem from the following table prepared by Lowrance: (4).

An Array of considerations influencing safety judgments

Risk assumed voluntarily	—	Risk borne involuntarily
Effect immediate	—	Effect delayed
No alternative available	—	Many alternatives available
Risk known with certainty	—	Risk not known
Exposure is an essential	—	Exposure is a luxury
Encountered occupationally	—	Encountered non-occupationally
Common hazard	—	"Dread hazard"
Affects average people	—	Affects especially sensitive people
Will be used as intended	—	Likely to be misused
Consequences reversible	—	Consequences irreversible

In this report we shall not attempt a complete classification of hazards and risks. Rather, by analysing the common elements of all the judgments of "acceptability", we shall clarify that idea, and thereby develop recommendations for improvements in society's management of risks.

2.3 Difficulties of Imagining Risks

2.3.1 Everyone deals with hazards on occasion, but they are quite different from ordinary situations where effects regularly follow causes. Behaviour that might at first sight appear incautious or even irrational might be comprehensible when the arbitrary and random occurrence of hazards is taken into account. Little is known of the judgments that people actually make in a real risk situation, but it cannot be a simple calculation of probabilities, costs and benefits. For instance, the greater weight given to occasional catastrophes over endemic small accidents must relate to some personal image of such events and their avoidance. A single-casualty accident can be imagined (however incorrectly) as one that could be avoided by prudence; those in a multiple-casualty disaster are, by contrast, doomed by the mere chance of being there at the wrong time.

2.3.2. We often handle irregular and unpredictable events in ordinary life; that is perhaps one of the attractions of gambling. But the skills and intuitions derived from such experience do not extend to the most important problems of risk assessment. For gamblers make choices between a few alternatives, each of them relatively likely; a horse at 100:1 is a "long shot". This is quite different from estimating risks whose odds lie between 10,000,000:1 and 10,000:1. When people play against odds in this range (as in the football pools), winners are almost always ascribed to luck rather than skill. Yet within this range is located a rough-and-ready distinction between those risks that are "acceptable" because they are no more likely than, say, being struck by lightning, and those that are not (or rather those whose probability is considered unacceptably high by the experts) (5). Of necessity, the experts in risk anslysis have refined their conception of risk. They need to be able to say: "This event should not happen more than once in a century", and still not be surprised, or consider their assessment falsified, if it happens a week later. But even if they do achieve such sophistication, they must eventually advise a lay public who undertands a "very unlikely" event to be one that really is not going to happen. One author has argued that actions involving fatal risks of 10^{-5} or less are in effect "safe" — because this is a level of risk that is commonly neglected by ordinary people. (6). Experts have a responsibility to prevent misunderstanding of their language by the public. They can make low probabilities more comprehensible, usually by translating from the risk for each action (as crossing a street), to that on a larger base (an individual's lifetime, or a larger population per year). But in this they must take care lest the new basis of measurement then becomes unreal to the person at risk; there is no easy way to relate the irregular and rare events back to ordinary human experience. For example, if some 20 nuclear reactors are expected to operate for 30 years, then a failure rate of "one in ten thousand reactor years" is difficult to express in terms of what people should expect to happen.

2.4 The Structure of Hazards

2.4.1. Although hazards are unusual experiences in many respects we can sometimes analyse their causes and estimate the likelihood of their realization. Then preventive or precautionary measures can be undertaken. Inquiries into accidents and disasters are organized around several sorts of question, and from these we can gain an insight into the different sorts of cause of a hazard. The focal point of an inquiry is the event (or events) that caused the harm. We ask (i) What happened? (ii) How did the event occur? (iii) What made it possible? (iv) What could or should have prevented it? and finally (v) Whose fault was it?

2.4.2 Any realized hazard (either one already occurred, or one imagined in design studies) can be analysed as the effect of the combination of the various sorts of causes indicated by the questions given above. We call this the "causal network". When these causes operate intermittently or very infrequently, the hazard itself will be less probable. The function of monitoring is to ensure that possible causes of an accident are kept isolated or at a low level, so that they do not connect. An example of a hazard with a complex but comprehensible causal network is a domestic gas explosion. What happens to cause harm is that gas of a certain minimum concentration accumulates in a confined space, to be ignited explosively when a naked light or spark is struck. *How* it happens may be the result of either a leak in the supply system, or someone inadvertently leaving on or knocking open a gas tap *and* someone or something striking a light. These are *initiating* events. Going further back in the causal network, we can explore the causes of each possible sort of leak; the remote causes might be decomposition of some channel or joint, or violent or excessive movement in the supporting structure, or a weakness in the appliance itself. The flame or spark may be made by a person unaware of the leak, or caused by an electrical failure. This latter cause would lead to a further network. For each possible chain of events leading to a gas explosion, there are circumstances we may call *enabling causes*. These could include ageing materials or faulty workmanship in both the gas and electrical supply system.
The failure of preventive measures may result from defective

monitoring. This would include neglect of inspection of pipes and fittings. Finally, the *fault* or blame for the explosion could lie with those who neglected or skimped "reasonable" established monitoring procedures; this would include carelessness by the user of an appliance. The causal analysis of a gas explosion may not stop there; there may be a *sequential hazard*, from the effects of the explosion. The most common of these would be fire, or structural collapse. Indeed, subsequent events may turn out to be even more harmful than the initiating event. The gas explosion at the Ronan Point high-rise flats, for example, not only blew out an external wall-panel, but also triggered a "progressive collapse" of the wall panels, first above, and then below the fated apartment. We also see from this example that our identification of the *primary hazard*, whether it be a gas leak, electricity sparking, or domino-type buildings, depends on our perception of causes and evaluations of effects. Indeed, the Ronan Point disaster led to the discovery of a previously unrecognised hazard; the inquiry discovered serious structural weaknesses in the load-bearing wall panelling of the building. It predicted that further "domino" accidents were more likely to result from high winds than from gas explosions. (7)

A full analysis of a hazard therefore includes possible sequential events along with the obvious primary one. Prudent management not only strives to reduce the intensity of the causes of the hazard, but also makes plans for the containment of the hazard should it be realized. These plans may include spatial features (barriers and separations) and also arrangements for the evacuation and care of people.

2.4.3 This analysis of the causal network of hazards helps us to appreciate how hazards come to be realized with a certain regularity, although infrequent and individually unpredictable. Relatively uncommon incidents must occur together or in a particular sequence (as a gas leak followed by a spark); under special circumstances (both in the same place); and where monitoring procedures are either ineffective or absent. Similar accidents may happen under similar enabling circumstances and with variations in initiating events. Families of accidents may show a remarkably regular pattern of occurrence, and to that extent may be predicted with considerable accuracy. A study of possible causes can often provide important clues which assist with preventive measures.

Safety engineers must be able to estimate the probability of occurrence of an event that is, of itself, unpredictable. Without this ability, they would not be able to choose between more and less serious hazards, nor to identify and tackle the most important causes. The essential tool for this is the causal network for the hazard. This may be represented either as a mathematical diagram, or as an informal awareness of particular danger points and their connexions. In this way, events that are individually unpredictable can be brought under control as a class.

2.4.4 We noticed that inadequate or ineffective monitoring can lead to an accident; it is well-known to those who regularly control hazards that monitoring procedures must be vigilantly observed. Thus, monitoring has a paradoxical feature: as an operation it must itself be monitored. Monitoring operations are often routine; they may be incompletely understood or inadequately performed by their operatives. Ultimately a personal concern by senior people is necessary if any monitoring is to be effective. But at the higher level there is no immediate test of competence, except perhaps a record of success — although even this may be inaccurate if the disaster does not show for many years (eg. thalidomide; asbestos). Hence, the quality of management, its technical competence, and its awareness of its moral responsibilities are important factors in the system which prevents tragedies from occurring.

The DC-10 air crash outside Paris in 1974, in which 346 lives were lost, is a classic case of a disaster caused by (on the most charitable view) ineffective monitoring. (8) The sequence of initiating events began when a baggage handler closed a cargo door at Orly Airport. He carried out with complete accuracy the drill he had been taught; unknown to him, the latching-and-locking system of the door failed to work. During flight, the door blew out, depressurizing the cargo hold. Pressurized air remaining in the passenger-cabin above collapsed the floor between cabin and hold, wrecking the flight controls carried on the underside of the floor.

The crucial enabling circumstance was the design decision to carry flight controls on a structure (cabin floor) possessing considerably less strength than the vital sections of the airframe (pressure hull, flying surfaces, etc) thus making it possible to lose the whole aircraft as a result of an intrinsically trivial mishap. This circumstance was much exacerbated by inadequate design of the cargo-door latching-and-locking systems.

These faults were exposed in ground pressurization testing and in a flight incident which almost led to catastrophe. The only modifications proposed were inadequate; in any case, they were not mandatory and had not been properly carried out on the aircraft which crashed in 1974. Those responsible for monitoring included the airline, the airframe manufacturer, the door sub-contractor and the US regulatory agencies. It is clear from this example that neither scientific findings nor externally-imposed monitoring routines can correct all hazard-producing situations. In the last resort, accurate analysis, discipline and commitment are all essential for the containment of hazards.

2.4.5 An examination of the disastrous train crash at an automatic level crossing at Hixon, Lancs, in 1968 showed (9) clearly how even the best-intentioned monitoring can fail when responsibility for the hazard is divided among several different agencies, each with its own wider responsibilities, routines and perceptions. For at this level-crossing there were lights, warning signs, and even a telephone for use by drivers. But these devices were not coordinated, and the instructions were in some respects ambiguous. And it had been no-one's responsibility to observe that a very long, slow-moving, low-loading trailer (of the sort that regularly carried heavy electrical generating equipment over the crossing) could occupy the crossing for a longer time than the warning system recognized. In this way, an event in the "incredible" class really could and did occur.

This last example shows us that a hazard is not simply an objective phenomenon perceived in the same way by all who are concerned with it. Rather, it is an intellectual construct, made by people each working within the confines of a particular social setting, each with their own way of perceiving the world. When we come to consider the problems of management of risks, we shall need to recall that it is only natural for the different parties to a hazard, including those who create it, those who control it, and those who experience it, to see it in different ways. An appreciation of this diversity of perception is of great importance for the development of means to achieve good management of risks.

Even the empirical data on hazards will be determined by the categories in which they are conceived. Hence (as we shall see in the next chapter), they can never be as "objective" as their mathematical form might suggest. Thus the human factor is present even at this most scientific part of the analysis.

3. THE ASSESSMENT OF RISKS

The assessment of risks can be tackled largely in a scientific manner. But there are limits to the strength of the inferences that can be drawn from statistical data and scientific theories. For this reason we have investigated in some detail the methodological problems of this approach, such as the identification of the hazard, the retrieval of sound data, and the appropriate use of mathematics. The observations we make are not so much intended to criticize existing practice as to note its limitations and indicate ways to its improvement.

3.1 Identifying the Hazard

We have seen (**2.4.5**) that a hazard is an intellectual construct; it is perceived quite differently by the different parties involved. Policy decisions about the hazard (including recommendations for the prevention of its recurrence) are influenced by this prior perception. It also affects the study of the hazard, down to the choice of salient phenomena and definition of data. To use a familiar example, we might distinguish between "*the* Flixborough disaster" and "*a* Flixborough disaster'.' The former was an event which, however unfortunate, was alleged to be extremely rare, and perhaps even unique. By contrast, "*a* Flixborough disaster" refers to a class of possible events at chemical plants of a type comparable to the one which actually blew up. The Committee of Inquiry was told to study *the* Flixborough disaster, and argued that it was virtually unique (see Appendix 3). However, the First Report of the Advisory Committee on Major Hazards makes it plain that "*a* Flixborough disaster" might well happen again.(10)

Hazards occur at every phase of operation of an industrial process; but those outside the immediate sphere of production have tended to be neglected by management, and their control is relegated to other agencies. The hazards of transport and waste disposal, for example, were not, until recently, seen as integral parts of the production process. The nuclear power industry in particular is now starting to realize that its original "linear" conception of the process was ill-conceived, and that a global, cyclical approach covering the whole "fuel cycle" is required. Since all nuclear waste cannot just be dispersed harmlessly, we must decide on the terms of the "Faustian bargain" (11) with future generations for the care of poisonous radioactive wastes. With the erosion of public confidence in waste-storage practices, the risks become less "acceptable" to society. Similarly, the regular international transportation of materials that are not only toxic but also militarily important, pose serious new security problems for governments (12). In retrospect, the Rasmussen Report on the internal safety of light-water reactor plants (13) may suffer a worse fate than that of being criticized as oversimplified and over-optimistic. It may simply be ignored, as dealing with one of the less troubling hazards of civil nuclear power.

3.2 Difficulties with Data

Accurate scientific estimates of risks are always difficult, and in some cases impossible. The very nature of the risk, particularly in low-risk situations, hinders the collection of the data necessary for the analysis. High-quality data, of the sort taken for granted in ordinary research, may well be an unattainable ideal.

3.2.1 When data are obtained in normal scientific practice, a measure of control is crucial. In a laboratory situation the experimenter tries to keep all but one of what he believes to be the relevant factors constant, so that by varying a cause he can unambiguously observe its effect. Clinical investigations rely on the "control group" for a proper test of a hypothesis by statistical methods. In field studies, a multiplicity of examples are studied in a disciplined way, thereby providing a basis for correlating possible causes with effects.

The scientist producing data for an assessment of risk has few, if any, of these means of control on the quality of his material. Hazards, inevitably involving relatively rare and unpredictable events, are peculiarly difficult to describe. In the case of an accident, reports are made after the event, often by witnesses who were confused and disturbed by it. Even when the investigator can use data that have already been collected and codified (as in the case of more common accidents and illnesses), these may be seriously distorted (for his purposes) by the circumstances of their collection. For example, under-reporting of industrial accidents may affect a quarter of accidents in some occupations, and three-quarters in others (14).

Similarly, the engineer seeking data for an assessment of risks in industrial plants faces a difficult task. There are as yet only limited data on the performance and failure rate of commonly used engineering equipment and on the interplay between equipment faults and operating routines and maintenance schedules. Obtaining a wide range of such data, in good quality, requires a large quality-assurance organization to be operating for a number of years, at a cost that will be a significant fraction of the contract price. In the absence of such a base, the investigator may have to start afresh and collect data from scratch. Even this will not always be easy, particularly if the data involve information that the firm wishes to keep secret. When Trade Unions share responsibility for industrial safety, their request for full access to data can raise very real issues of participation and power within the firm; safety may then become the focus of a political struggle.

Rates of occurrence of relatively infrequent events can be expressed in many ways. The relative risks of different modes of transport look rather different, depending on which base is being used (see Appendix 1). Another good example is that discussed by Kletz (15). Storage tanks for hydro-carbons explode (with a man on top) with a frequency of only once in every 8×10^8 working hours. As it stands, this is considered a "negligible" risk to employees; but when it is realized that the explosion usually occurs when a man is performing some operation on the tank, the risk relative to his work (rather than to his location) is some 20 times higher. It then becomes salient in comparison with other related risks, and can no longer be considered either "negligible" or "acceptable", and safety engineers then take action to reduce it.

The very terms in which the data are defined may well be a matter of choice and controversy; and different conclusions may be drawn from different sorts of data. In the field of industrial accidents, a convenient measure is the "severity threshold" of a "three-day disability". One recent study showed that such accidents have increased in recent years; and these figures correlate significantly with the budget of the Factory Inspectorate. However, in the same period the fatalities also decreased significantly. Since the victim of an accident has a choice over whether to claim a three-day injury, but none over his death, it is fair to conclude that subjective factors have influenced the particular statistics on injuries (16).

3.2.2 The difficulties of obtaining data on hazardous events are even greater when the events themselves go undetected; only to be deduced from long-delayed illnesses. Many pollution hazards are of this sort. It is quite impossible to maintain an environment free of all traces of toxic artificial substances; and the only way to be sure that a particular pollutant is dangerous is to establish a link with an illness. However, there are often major obstacles in the path of anyone who tries, as we see from the following examples. With PVC, the extreme rarity of the induced complaint (a cancer) in the population as a whole was partly responsible for the long delay in identification. Cyclamates seem to come on and off the American danger list as new evidence becomes available and is then refuted. And DDT, in spite of its indubitably damaging effects on the non-human environment, has been of incalculable benefit and has never (it is claimed) directly caused a human fatality (17).

Even when an illness is acute and quite localized in occurrence, sufferers may be neglected or wrongly diagnosed by medical men; thus the symptoms of mercury poisoning in the victims of the Minimata tragedy in Japan (18) were for years simply dismissed by their doctors. Closer to home, the term "asbestosis" has been commonly used to describe fatal cancer caused by asbestos (see Appendix 3). It took a "tremendous fight" (19) for Nancy Tait to get the true cause of her husband's death — mesothelioma — on his death certificate; and at the inquest the coroner refused to call evidence of asbestos fibres in the lungs (20). Legal liabilities may be affected by such descriptions, and mortality statistics can be affected by such considerations.

A medical man can so easily miss the signals which would show that an ailment is not one of those many mild, vague, and perhaps partly psychogenic complaints for which there is as yet no cure and no known cause. Medical treatment is usually structured around the place of residence rather than around the workplace where the patient may be exposed to hazards; hence correlations between disease and exposure may be lost altogether or buried in the records system. And where an institution does provide medical examination and treatment, those in charge have opportunities for distorting or suppressing medical data that would be costly or embarassing to them. It is easy, for example, to bury data on the significant exposure of a few members of the workforce beneath figures on the health of all workers in the factory, including inspectors, cleaners and administrative staff. It is therefore incumbent on those who use "illness data" in quantifying risks and hazards, that they scrutinize them carefully to make sure they are sound, relevant and free from distortion.

3.2.3 Critically difficult problems of data collection arise when the events in question are speculative, as in the case of designing a new type of industrial plant or structure, or estimating the risk to future generations from effluents currently released. To the extent that the causal network of the hazard is similar to those already studied, existing data (for example on the non-catastrophic failure of components) can be used with some confidence. But in a radically new situation where the environment of hitherto standard components may create new and unpredictable hazards, past experience is less reliable. Data from increasingly more sophisticated and accurate simulations of operational conditions can be prepared, but only after a lengthy and expensive R & D programme. And when the innovation constitutes a "major hazard" in which tolerated projected failure rates are very low, the problems of securing adequate data on the weaknesses of untried components become very severe indeed. Comparing this case of hypothetical, very rare disaster, with that of the common minor accident, we see how all the problems of hazard analysis, starting with data, are here both more demanding and more difficult.

3.3 A Prudent Policy for Improbable Hazards

All analyses aim to reduce hazards so far as possible. But there is only one way to reduce to zero the probability of an accident, and that is to remove its causes. As engineers say, the only perfectly safe aeroplane is the one that stays on the ground, either in still air on a disused airfield or in a locked hanger. Risks may be reduced to a point where they are "acceptable" in some sense or other; but so long as the causes are present, the effects will, to a degree, be there also.

There is a borderline area, which has now become important in public policy on hazards; those hazards whose likelihood of occurrence is described as "negligible", "astronomically small," or "incredible", but whose consequences can be very serious. These descriptive terms can be related to probabilities (on some appropriate base) of around 10^{-6}, around 10^{-9} and around 10^{-12}. Such numbers can be useful as cut-off points in quantitative design studies for catastrophic accidents. But if they are taken literally as exact limits of permitted risks, very serious errors can occur. The calculation of these very small numbers can be an uncertain business itself; when they are used in argument, they should be carefully scrutinized to see that they possess real meaning (see Appendix 7). For in the case of new and complex installations, those very small probabilities cannot be derived from trial-and-error experience. Arguments about such hazards lie in what has been well described as "the domain of hypotheticality" as they are "necessarily and ultimately inconclusive" (21). An impartial confirmation of this status is provided by insurance companies. When (as in the case of civil nuclear power, in the UK and in the USA) an industry does not or cannot obtain full cover on the insurance market, it may be fairly inferred that its risks are other than completely calculable.

The case of hazards of ammunition (Appendix 4) serves in many ways as a paradigm for this situation. Only an independent, rigorous and dedicated quality-assurance organization is good enough for the control of those hazards. No possible source of an explosion is discounted. In reviewing arrangements for monitoring and containment, the motto is adapted from the classic "everything that is not forbidden is compulsory", to read "everything that is not impossible (for people to do incorrectly) is inevitable". The strong possibility of a highly unlikely contingency is perhaps best shown by the example of the risk of a terrorist attack on a nuclear power station. This risk was quite neglected for the first 20 years of the civil nuclear power programme.

The proper control of risks will then depend on taking *every* sort of risk seriously, and deferring action on none except those which are publicly shown, through frequent periodic re-assessments, to be (truly) "incredible". Until recently it was enough to have a scientist assuring the public, on behalf of some industrial interest, that their worries were groundless. But the credibility of industrial safety experts has been somewhat impaired by disasters in several science-based industries. Also, scientists in the American nuclear industry have been the target of criticisms of consumers' advocates on the grounds of alleged concealment of hazards (22). From now on, experts representing industrial interest will have to face the difficult task of convincing a suspicious public that each particular improbable hazard (existing or proposed) is so very unlikely that it may fairly be acceptable. This might be a salutary experience for them, and the resulting dialogue might well lead to fairer decisions and more effective controls.

3.4 Summary

In this chapter we have shown that the assessment of risks, while being scientific in its methods, cannot by itself provide the answers to policy questions. Indeed, the special difficulties of analysis of hazards require scientists who are particularly skilled in their methods, and aware of the broader context of their work. We shall return to this point in Chapter 7; but first we must review the less "objective" aspects of the analysis of risks.

4. THE EVALUATION OF RISKS

4.1 Costs and Benefits

When we consciously take risks in ordinary life, we may pause to ask "is it worth it, to do it this way?" The question may involve an aeroplane journey, overtaking a car, or crossing a street at an unprotected place. There is an implicit estimation of the benefits of that particular action in comparison to alternatives, and of the risk it entails. All this is done very informally, with no attempt at precise calculation; we are guided by our own and others' experience, and usually cannot delay for a long examination of the problem.

When decisions are made that involve the creation of significant hazards, casual procedures are inadequate. If detailed examination of all costs and benefits of the project is to be attempted, then risks must be carefully analysed. But how is this to be done?

In principle, it would seem straightforward: to assess the probability of a realized hazard, and then to estimate the costs of the consequences. Indeed, it has seemed that some such objective, impersonal evaluation of risks is necessary for there to be fair decisions about hazardous technological innovations.

The hopes of scientific measures of social and personal costs have largely remained unrealized. And the prospects for achieving them are acknowledged to be remote. Here we will briefly examine one case of special relevance to risks, in which precise objective evaluation has turned out to be impossible. This does not mean that we abandon hope for fairness in decisions on risk; but, as we shall see later (Chapter 6), its basis will lie not so much in scientific facts as in the procedures governing dialogue.

4.2 The Limits of Measurement

At one time it seemed that the techniques of "cost-benefit analysis" would enable experts to evaluate social and technical policies in terms of a common unit — a form of "social money". The community would then be able to balance these, and all sides would concur in a rational judgment. The hopes of such an objective solution to essentially political problems were soon frustrated; and cost-benefit analysis is now restricted to being at most a rough guide to policy. But the problem remains: how to establish the basis of a dialogue between sides with opposing values and perceptions of the hazard. We shall later (Chapter 6) show that final, objective solutions to the evaluation of risks are neither possible nor necessary. But we should briefly indicate the severe difficulties that affect any attempted measurement of consequences, so that the case for an alternative approach is well established.

4.3 Estimated Costs — The Example of Human Life.

The prospect of a universal measure of the cost of harm is alluring. But a look at the important example of human life gives some idea of the limits of legitimate inquiry. Human fatality furnishes the unit of harm in various applications, with personal injuries being costed as some proportion of the cost of a death.

We all make some sorts of reckoning, and allow society to make others for us, in which we implicitly assign bounds to the value of a life in a risk situation with a significant probability of death. But since life is not merely another commodity, we should not be surprised to find a variety of approaches to its value. In fact, estimating the value of a human life turns out to be highly complex and contradictory. In analysing the approaches to the value of a life, we may imagine three "interests" in the situation: the potential victim, the maker of the hazard (who may be the victim himself), and others (a third party, or society at large).

4.3.1 The "subjective" valuation of life, by the potential victim himself is of considerable importance for any attempt at a theoretically "fair" imposition of risk. Society must provide some commensurate compensation if there is to be any justification in imposing risks on particular groups of people for the sake of a general welfare. Some interesting studies have been based on the behaviour of people in hazardous situations. A simple example is that of the choice to do something that saves time, at the cost of a known risk to harm: crossing the street away from a protected point. The person is assumed to be making an implicit calculation of costs and benefits, and to act when the balance is favourable or equal. If the benefits can be given a cash value (for time saved), and the probability is known, then an upper bound on the "subjective" value of the life can be derived from the inequality "Benefit > Probability x Harm". (23)

Some earlier studies along "behaviouristic" lines seemed to offer the hope of discovering several mathematical regularities in subjective evaluations of life; thus C. Starr obtained the relation "the acceptability of a risk (of a fatal accident) seems to be crudely proportional to the third power of the benefits felt or imagined". (24) Unfortunately, later, more extended research has not confirmed these regularities (25). Moreover, at higher levels of risk), where a person must seriously ponder his or her fate, the simple algebra of costs and benefits is inadequate.

4.3.2 It might seem that the explicit practices of certain social institutions would provide a better approach to evaluating the value of a life, than the implicit reckoning of individuals. In the courts and tribunals that set compensation for loss, lives and limbs are valued as a matter of routine. But here too, it proves impossible to escape from methodological difficulties. In assessing compensation payments, the judgment does not avoid considerations of fault.

Compensation payments are, in addition, strongly influenced by the estimates of lost earnings. In this way the retrospective value of a lost life is more for a rich man than for a poor one, and more for either than for his wife.

Using materials and techniques from economics, some analysts have attempted to provide a uniform, objective measure of the various sorts of lives at risk. But here, decisions must be made on the implied transaction involving the life. Do we take earning power, representing the risk to dependents? Or do we take a measure of a lost contribution to society at large, which might be the difference between production and consumption over some years? This is a very theoretical concept, involving many special assumptions.

In view of this necessary and inevitable artificiality of computations of the economic-social value of a life, it is hardly surprising that occasionally paradoxical, and even ludicrous, results occur. Looked at in one way, it appears only reasonable to observe that an old-age pensioner is certainly an economic liability on society and a woman probably so. But then if the economic-social cost of a fatality to such a person is registered in the calculations as negligible, or even as a benefit, something is wrong. Risk analysts must then need to invoke arguments outside their technical competence, to avoid the conclusion that traffic on roads performs a social function in culling the human population. (26).

4.3.3 We have not yet mentioned the third "interest" in the hazard — those individuals that produce the risks that are endured by others. The owner or operator of a hazardous system may have to balance the direct cost of reducing a risk against the low probability of cost (or penalty) for letting that risk continue. The immediate pressures on the responsible individuals are to "accept" the risk which is thus imposed on others. Where the strengths of the different "interests" are unequal the effective balance of cost and benefit is then heavily tilted against the potential victims. The history of occupational health and safety in this country provides countless examples of this imbalance. Even now, the risk to workers in some parts of industry are far higher than the levels considered acceptable elsewhere. (27) This state of affairs is sometimes rationalized by the assignment of fault to those enduring the risk. A startling example of blaming the victim for his misfortune was recently produced by the Asbestos Information Committee (28). Answering its own question "Can Asbestos Be Used Safely?", it answers: "Yes. Any risk comes from careless working on asbestos products which can cause you to breathe too much asbestos dust". (See Appendix 5).

4.4 The Calculation of Benefits

If the costs of a realized hazard (as measured by the value of a life) are difficult to calculate exactly, the benefits are even more so.

In the very simple case of crossing a road, one may estimate the value of time saved or foregone: but we have seen that where risks are part of an unalterable environment (as they can be at work), the achievement (and calculation) of benefits is far from straightforward. When cost-benefit analysis is applied for large-scale decisions on technology policy, as in cases of transport and energy supply, the social and environmental dimensions of cost and benefit, all necessarily calculated for a hypothetical future, become very inexact indeed.

The classic failure in the application of cost-benefit analysis, the Roskill Commission's Report on the site of the proposed Third London airport (29), may have foundered on just this point of inexactness. When its highly speculative methods yielded only some £200 million differences in the "total resources cost" between the leading alternatives estimated at around £4,300 million, its authors might have made clear the 5% was not a significant difference for this type of study, and that the choice was therefore purely political. Instead, they made a firm policy recommendation, saw their small costs-difference whittled away by special arguments (it would be largely accounted for by a 5 minute average difference in travelling times), and finally witnessed the political debate that decided the issue against their recommendation.

The benefits of new technologies are often very real, in spite of being incapable of exact or certain prediction. It would be unfair to proposers of innovation, and harmful to society, if in every debate the immediate and calculable risks were allowed to outweigh the possible social benefits. (30) But if the experts on the side of innovation become discredited because of previous failings, the opponents of change will convince the public every time. The handling of this very difficult problem, which actually involves several distinct phases of innovation has been discussed in the Council's earlier report on the monitoring of technologies (31).

4.5 Using Inexact Estimates of Consequences of Hazards

Our critical analysis of the estimates of the consequences of hazards could be thought to be leading towards the negative conclusion that meaningful estimates are impossible. The corollary would be that reason has little part to play in decisions on acceptability — might establishes what is right. In fact the situation is not so desperate. A legitimate and careful use of estimates is essential for debates, though the slipshod or partisan application of scientific techniques can lead to a discrediting of experts, of the dialogue process, and of science itself.

4.5.1 While some branches of the physical sciences are able to achieve spectacular accuracies of measurement (attaining parts per billion in some cases), high precision is not common in the life sciences, and in the behavoural sciences it is rare indeed. Moreover, in statements involving probabilities, a simple number (usually adequate for describing physical measurements) may be quite misleading. Conclusions about the significance of the association of possible causes and effects must have a "confidence limit". This may be given as a percentage, roughly equivalent to the betting odds that the conclusion is correct. Also, when inexact data and guessed parameters are fed into a mathematical computation, the conclusions may be very inexact indeed. On occasion we should speak of a "magnitude-band", expressed as, say, "to within a factor of 10". (see Appendix 7). We possess no convenient symbolic notation for this essential aspect of clear description and thought on quantitative statements, and this may be one reason why so little attention is paid to inexactness. But to quote a precise number to the public without any indication of its inherent "spread" is really unscientific, and may be as misleading in its effects as stating one that is simply wrong.

4.5.2 By keeping in mind the essential inexactness of estimates of consequences, we can avoid some troubling problems that can afflict the analysis of risk. We noticed earlier (**4.3.2**) that if a minimal or negative "economic value" is ascribed to the life of an old person, the logical consequence would be to use the roads for culling the population. But if all the highly inexact quantities in such calculations were cited and reckoned with the appropriate magnitude-band, then such socially absurd conclusions would be less likely to emerge. Also, mathematical arguments with only spurious content would be more easily recognized as such.

On the deeper problem of evaluating costs, the only genuine solution is to abandon the attempt at a uniform measure of life, and treat the various dimensions of the costs separately. In mathematical terms, we would say that any measure in this field involving more than one term must be a vector. There may then arise a situation where estimates of such mixed costs must be compared, and predicted lost lives traded against other costs and benefits. This can be a macabre operation, but it is honest and above-board, so long as the lives at stake are there to be seen, and not lost in some aggregated monetary total. Also, this is where politics enters the decision-making process legitimately and explicitly; there can be no pretence that scientific expertise can balance the different dimensions of a total cost.

Those who make decisions on health and medical care have for a long time coped with this same problem. There they accept the impossibility of a completely scientific and fair solution to the problems of priorities; they operate within the financial and structural bounds that society imposes, and the essentially insoluble moral dilemmas of life and death are treated with respect (32). Cost-effectiveness techniques can then legitimately be applied to the policies that implement the basic decisions on priorities among lives.

In cases of large-scale technological decisions, skill and sobriety in the use of scientific methods will be of the utmost importance. It will not always be easy for scientists to give magnitude-bands for those quantities that can be estimated, to give confidence-limits for probabilities of events, and to confess ignorance when that is appropriate. Yet to do otherwise would be to fail in their duty, and to discredit the principles of rational debate.

4.6 Our analysis in this chapter has been a case study on the problem of giving quantitative measures of qualitative judgments. We cannot omit considerations of value from decisions on hazards;

but neither can we legitimately reduce them to a simple number. One way around this dilemma is to appreciate the inexactness in all measures, and to learn to cope with the high degree of inexactness in this particular case. But the human factor in risks is not completely encompassed by that approach. We shall see in the following chapters how judgments of acceptability of risk, and of fairness of decisions, depend upon the personal situations of the parties to a dialogue on a hazard. Appreciating this, we find that rational discussion of risk does not depend exclusively upon successful qualification, though as always, scientific methods are an important part of any effective study.

5. JUDGMENTS OF ACCEPTABILITY

5.1 The Variability of "Acceptable" Levels of Risk

If the seriousness of a perceived risk were a simple matter of estimation of probabilities, costs and benefits, then we would find roughly similar levels of acceptability of the various risks that people encounter. But there is no such uniformity in practice. Some risks resulting from personal indulgence and convenience (as from legal drugs and driving at high speeds) are allowed to persist at quite disproportionate levels. Also, manual workers and their families endure excessive risk in employment and at home (33). These and other disparities might seem to be evidence of widespread "irrationality", leaving the scientific experts as the only ones competent to make judgments of acceptability. Indeed, some who wish to counteract popular hostility to particular suspect industries (as civil nuclear power) have even called for research into the ulterior motives of those who demand very low levels of risk in those cases. (34).

As we shall see (chapter 6) leaving the judgments of acceptability to the experts will not necessarily guarantee either fair decisions on risks, or effective control of them. We do better to appreciate the actual varieties of judgments of acceptability and to make that the starting point of our analysis. We shall find that "the acceptability of risk" is not a simple idea capable of being reduced to some uniform measure. Several sorts of "acceptability", very different among themselves, can be invoked in decisions on the creation or maintenance of risks. Most important, a risk may be "acceptable" in practice even when it is manifestly unjust to those enduring it. Hence we shall use the term "acceptable" with care, avoiding the confusion of what is, with what ought to be.

5.2 Varieties of Perceived Hazards

The table of characteristics of hazards that we borrowed from Lowrance (see **2.2**) shows how many considerations can be involved in the judgment of their acceptability. Here we shall provide a few examples, chosen mainly to illustrate different sorts of acceptance (or rejection) that may be involved.

5.2.1 Some hazards are accepted voluntarily, even when the risk is very high. At one extreme, we may say that the risk is "embraced" when it is an integral part of the challenge in a hazardous sport, such as pot-holing or motor-racing. Facing possible injury or death is a skill to be mastered along with physical techniques, and the sport would not be the same without it. The personal attitude is different in emergencies, where the risk may be said to be "defied" in the course of a response to a call for help. Rescue operations are the prime example of such hazards. In them, the balance of costs and benefits may be very different from that which prevails in ordinary life, as we can see from the lengths to which kind-hearted people will go in retrieving lost or trapped animals. It is, of course, impossible to describe in general, just how people perceive the risk to themselves in each case. Psychological studies have not yet succeeded in isolating reliable indications of the structures of perceived risks or the motivations of those who undertake them (35). Certainly, in ordinary hazardous situations there is a tendency to "dismiss" the risk, with an attitude that "it can't happen to me". This is perhaps psychologically necessary in cases of repeated exposure, voluntary or otherwise.

5.2.2 When a serious hazard is encountered involuntarily, acceptance may extend only to a much lower level of risk than otherwise. When, in addition, the sufferer feels impotent in the face of danger, tolerance is further reduced. Accidents in trains seem peculiarly unacceptable, perhaps more so than accidents in aeroplanes, where rightly or wrongly the passengers are generally considered to have taken the risk on themselves for the sake of the extra benefit of the time saved. In underground tube-trains only absolute safety seems to be good enough: perhaps the enclosed environment exerts a strong psychological influence. The stark terror of impotence in the face of impending destruction is an important part of the evaluation of such risks. For a strong contrast, we notice how the illusion of control by a driver in a private motor car who is under the influence of alcohol makes very high risk levels acceptable to him.

5.2.3 Perhaps the most difficult class of hazards for judgments of "acceptability" are those called "major hazards". These are defined by a low probability of realization, combined with the likelihood of very great harm if the hazard is realized. In this case, the intuitions derived from ordinary experience provide little help in conceiving the hazard; and the experts themselves may well disagree even over probabilities. (see para. **3.3**). When harm is liable to be inflicted on people who have neither any conceivable power to avert it, nor any respnosibility for its occurrence, the judgment of the "acceptability" of the risk is at its most difficult. Chemicals with teratogenic or genetic effects produce such a hazard; and in this respect, nuclear pwoer is a particularly severe "major hazard". Also catastrophes that could cause a temporary or permanent breakdown in civil order have a peculiar horror of their own. There are indeed some substances and processes that are popularly considered "absolutely unacceptable" because of their extreme lethal powers. But there is very little that is truly absolute in this world. The security precautions at biological-warfare laboratories were considered adequate for the containment of virulent pathogens in spite of their being, in the last resort, established by fallible human beings. And the risk of a thermonuclear holocaust is one which most people have learnt to "tolerate" in some sense or other.

5.2.4 This variety of perceptions may well be a cause of irritation to an expert on risk assessment. Policy decisions would be so much easier if a purely quantitative analysis were enough or nearly so. The subjective perceptions of risk can have enormous political importance, possibly to the extent of distorting priorities in programmes for coping with the real risks that society encounters. (This problem is most noticeable in medicine). There is occasionally a temptation on the part of the expert in technological risks to throw in a couple of extra orders of magnitude of restriction on the emotionally unacceptable hazards (36) and then perhaps to be annoyed when even this does not render them acceptable to critics. But in every hazard the various interests naturally and legitimately bring their own valuations and perceptions to it. Even the scientists do not always reach an unambiguous, conclusive assessment of the severity of risks. To suppose that people can or should have the same perception of risk is naive, and not useful for understanding or improving the way society actually copes with risks.

5.3 Risk Levels and Judgments of Acceptability

The main criterion for distinguishing between risks that are "acceptable" and those that are not is their "severity", a quality compounded of probability and harm. But there is no simple scale with a single cut-off point above which hazards are prohibited, and below which they pass unchallenged. Subjective perceptions of risk, as we have just seen, often outweigh their objective, partly quantifiable, aspects. If we keep in mind the varieties of interests in a hazard, with each individual having his or her own perception and evaluation, we can explain to some extent the apparent anomalies in the ways people deal with risks. Here we will review some different sorts of "acceptability" that are invoked in judgments on risks. Our analysis will be a slight refinement on that of Knox (37) adapted to the policy problems we discuss in this Report.

5.3.1 The extreme case of an "accepted" risk is one that is *totally unknown*, whose existence is quite unsuspected by all concerned. But since every activity or substance carries some risk of harm, we may say that this class of "unsuspected" risks is only an extreme case of those that are *ignored as negligible*. Here, either the probability of realising the risk, or its harm, or both jointly, are not reckoned to be sufficiently serious to influence policy decisions. But any identification of a hazard as "negligible" must be tentative (see para. **3.1**). Thus in England we do not design buildings to withstand major earthquakes, nor did we protect them against high winds (100 miles an hour or more) until after the Ferrybridge cooling-towers collapsed in November 1965. Likewise we did not consider the possibility of concerted attacks by terrorists when designing domestic and industrial installations. It is a salutary reminder of the provisional character of all risk assessments that in the case of nuclear plants that particular risk is now seen as far from negligible — several installations have already been sabotaged (38).

5.3.2 In some situations, risks are relatively easy to quantify and to relate to possible causes; industrial hazards are one such. The risk-analyst may discover hazards that, although quite real, present small risks by comparison to others in the same environment. He would consider such risks *less salient* and naturally, with his limited budget, he would postpone dealing with them until more serious risks had been tackled. As a strategy this is sensible; though what is "less salient" may depend on one's point of view. Also, if a set of identified small hazards relate synergistically to one another, their aggregated risk may approach or exceed that of any single other hazard; in which case, postponing treatment may not be so easily justified. This problem has arisen in connexion with radiation risks, where the multiplicity of substances, pathways and effects requires that single calculations be used with caution (39).

5.3.3 In some instances, a risk may be obvious and salient, and yet it seems to escape a really serious effort for its reduction. We may consider such a risk to be *socially permitted* in spite of any propaganda to the contrary. A sign of such "permission" is the lack of intense concern (on the part of those who control the risk) to prevent the recurrence of any particular realization of the hazard. Until recently, industrial hazards were firmly in this class, though there is now a welcome change. By contrast, accidents on roads, and in particular to pedestrians, seem to be "socially permitted" so long as the number of fatalities is not seen to rise dramatically. The absence of intensive experiments in speed control and other safety measures is a sure sign of a relatively low level of social concern. Only after a major road disaster, especially one involving a public-service vehicle, will there be an official inquiry of the sort that is routine for all rail accidents, however minor. It seems that the various benefits to the large and influential groups of those who drive on roads, as well as the associated industries, are sufficient to make the risks "socially permitted".

5.3.4 It is perhaps a subtle shift to those risks that are *considered unavoidable*. But the difference lies in the attitude to the future: whether the risk is seriously expected to be reduced. In industrial hazards and pollution, the formula "best practicable means of control" is double-edged. It may serve as a cover for complacency about the present (in which case the risk is "socially permitted"); but (in view of the constant changes in technique and equipment) it can serve as a cutting edge for steady improvement. Such risks may be described as *temporarily tolerated;* the case of the hazards of air transport is a good illustration of this (Appendix 1).
We shall see (Chapter 6) that the expected future of a risk is an important aspect of its present fairness.

5.3.5 There are important hazards where, to a dismaying extent, those exposed appear to accept the danger to themselves almost willingly. Very many people spend their working lives in the presence of high risks: for them risk may be endured even though it is simultaneously *irreducible and intolerable*. Such people are not like the daredevil driver out for thrills who does, after all, play a game in courting disaster. Rather, they may appear peculiarly apathetic in the face of avoidable danger, or even irrationally careless of life and limb. In the workplace this may stem from a sense of impotence; since the risks cannot be reduced the victim abstracts himself from the problem. The reduction of risks requires a management prepared to invest financial and organizational resources, a workforce willing and capable of taking a positive attitude to their own health and safety, and a determined and effective inspectorate. Where such attitudes prevail, and capital is available, it can pay for itself in every way. But only too often it becomes company policy to reduce risks to workers only *after* a major public scandal, or where their hand has been forced by a determined and articulate workforce.

5.3.6 Finally, we should note the only sort of risk that is truly "acceptable" in the ethical sense: the risk that is judged *worthwhile* (in some estimation of costs and benefits), and is incurred by a deliberate choice made *by its potential victims* in preference to feasible alternatives. The previous example reminds us that not every "accepted" risk is truly "worthwhile" in proving beneficial for the person enduring it. Also, we shall later (Chapter 6) examine the policy of requiring all socially imposed risks to be "worthwhile" to those who bear them. Although this might seem to be the only fair criterion for the creation of large-scale risks, we shall see that it is fraught with difficulties of its own.

5.3.7 Evidently, there is considerable variety in the situations and values that may be involved in a judgment of whether a risk is "acceptable". The risk may appear "negligible", "less salient", "socially permitted", "unavoidable", "temporarily tolerated", "irreducible and intolerable", or finally "worthwhile". Only in the last case is there any strong correlation between the acceptability of a risk in practice and its fairness in principle. Indeed, risks are quite commonly "accepted" in fact, while being "irreducible and intolerable" to those enduring them. Our discussion of policy and ethics will therefore not use the concept of "acceptability"; instead we shall explore the "effectiveness" of controls and the "fairness" of decisions. We can now also appreciate the limitations of the quantitative approach to the assessment of evaluation of risks, as discussed in Chapters 3 and 4. Where "acceptability" depends too much on the context and prospects of a risk, scientific estimates tend to become evidence in a complex judgment (perhaps crucial on occasion), rather than facts that entail a policy decision. Having established this structure for the problems of acceptability of risks, we can now proceed to consider ethical problems and practical tasks in Chapters 6 and 7.

6. FAIR DECISIONS ON RISKS

6.1 The Ethical Problem

In our discussion of the logic of the judgments whereby the "acceptability" of risks is decided, we have reviewed a series of difficulties in a scientific style of approach. The tasks of securing precise and objective judgments on risks become more difficult as one moves away from simple assessments of probabilities into the area of values, perceptions and criteria of acceptability. Indeed, on that last point we observed such a wide variety that the usefulness of the idea of "acceptability" is seriously reduced.

We should not conclude from this that there is no possibility of consensus on the fairness of risks. We can see that the ethical problems of risk must be investigated if we are to establish a basis for real fairness of decision and, consequently, effective control of risk.

The ethics of risk now has direct practical implications. Although many traditional hazards have been reduced or abolished, modern technology has now produced some potentially catastrophic ones. Populations exposed to such hazards are becoming aware of the possibility of questioning their acceptability. For a hazard to be allowed to persist, there should be some sort of consensus, where the tangible benefits of its presence are seen to outweigh the likely costs, either of its possible realization or of its elimination. But when the benefits do not accrue equally to all sections of the community, ethical problems are encountered. The question then becomes: *under what conditions, if any, is someone in society entitled to impose a risk on someone else on behalf of a supposed benefit to yet others?*

In a society such as ours, with a long tradition of valuing both the individual himself or herself and the notion of democratic rule, two strands of ethical thought can be brought to bear on the problem of when a risk is fair. One strand will see fairness in terms of the rights of individuals; the other in terms of what can be competently judged as best for the community as a whole in the "public interest". We may test the adequacy of these approaches against the practicability of using them to guide our judgments of fairness of risks.

6.2 The "Individualistic" Approach

In our previous discussions of the varieties of risk, we touched on the form an "individualistic" approach would take. In that perspective, only those risks judged "worthwhile" for *and by* the exposed persons could truly be said to be "acceptable" or fair in an ethical sense. The judgment of a risk being "worthwhile" presupposes the following: that each person undergoing the risk has access to the best available knowledge of its causes, its probability of occurrence and its likely consequences; that he can assess the probable costs and benefits of the given risk in comparison to those arising from other possible courses of action; and that he is free to choose whether he will expose himself to the risk or not.

But difficulties appear when we ask about those who do not accept the given risks as worthwhile to themselves. Should every new development in technology be subject to every individual's veto? Clearly such an arrangement would not work. We might think also of the question of compensation, either for the toleration of an irreducible hazard, or for the inconvenience of being removed to a safer situation. But the problem of fairness arises again: how is the level of compensation to be decided, and by whom? We have seen that one cannot "value" a risked life for its possessor as one can value a house for its occupant. Should we therefore, in all consistency, allow objectors to set their price for tolerating either the risks or some remedial measures? Only in this way could each person accepting the risk, or avoiding it, be able to consider himself treated with complete fairness. Also, since any experienced person knows how difficult it is in practice to get full compensation, the aggrieved party could insist further on procedures that ensured his full satisfaction. By all these means to ensure fairness we could find that society could be held up to ransom on any and every proposed innovation, the most unreasonable objector benefiting most handsomely.

Thus the "individualistic" approach can be seen to require too many unrealistic conditions and to give individuals the power to veto important developments, regardless of the reasonableness of their fears, or to extort exorbitant compensation. But, lest we dismiss this approach out of hand, we should consider how we ourselves would react when threatened by hazard, and we would certainly be the first to endorse this procedure. So we can use the individualistic approach as a Utopian ideal; while it is impracticable, it does give the greatest possible recognition to the individual's aspiration for the right to make his own choices on matters concerning his fate.

6.3 The "Public Interest" Alternative

By contrast, the "public interest" alternative is based primarily on the needs of society. Here, the assessment of risks, (along with socially-imposed costs) is to be handled more objectively. In each case of risk, a calculation is to be made of the overall cost and benefits to society of the policy which involves the risk in question. The fairness of the judgment lies in its being an evaluation on behalf of society as a whole. But its legitimacy can be ensured only by entrusting it to experts who, as public officials, are constitutionally appropriate for making such evaluations on society's behalf. The attractiveness of this approach lies in its implied assurance that the weighing of social costs and benefits can be removed from the scene of political conflict.

But this approach also has its own difficulties. All our previous discussion has shown the futility of attempting to create a precise mathematical science of risks, costs and benefits. Data gathering in this field is far from straight forward. Which groups of people at risk do you include? How many of the possible causes of a hazard should be reckoned with? And so on. Applying mathematical models to infrequently occurring phenomena, particularly where very large hazards are incurred, is largely a matter of judgment. And quantified social costs and benefits are meaningful only within broad bands of magnitude. Thus there is simply no prospect of ever achieving conclusive and convincing precise mathematical arguments on the fairness of risks.

Futhermore, there are ways in which judgment by technocracy can make the achievement of a dialogue between the conflicting parties more difficult. There is no common language between the inexpert members of the public who are worried, and the experts who interpret the hazards. There is no untutored common-sense image of a hazard, from which ordinary people can derive some feeling of control over it, and the consequent sense of security. Hence, technological hazards once discovered, have an especially dread quality. Nor can they be passed on (unlike many commercial costs) and may not even be mitigated (unlike in some environmental and amenity ones). As a result of this confluence of factors, those protesting may well develop an intensity of emotion that seems misguided or irrational to those (in positions of safety) who can approach the problem in a detached scientific manner. Those citizens and their advisors who have trained themselves in the technicalities of the hazard, are liable to be scorned by the qualified experts in spite of possessing full competence on the problem in question. And finally the choice of the right sort of "acceptability" is a crucial policy act, on which debate may rage equally with that on the (partly) "objective" assessment of the risk in question.

6.4 The Dilemma

Thus, no single, static conception of the criteria can yield an answer to the question "What is a fair risk?" We seem to face a dilemma between the disadvantage of individual veto or societal fiat. Rather than continuing to demand an answer to the question whether a risk is fair *in itself*, we should redirect our attention to the ways that risks come about and are controlled. That is, we should focus on the *procedures* by which decisions are taken on the creation or persistence of risks, and ask whether these procedures are fair. In doing so we would be following some fruitful researches into other areas of social ethics, where concepts of justice and fairness are central (40). Of course, this is necessarily a relative concept of fairness, since inequalities in the skills of manipulation of formally fair procedures are an important part of the means of maintenance of inequalities of wealth and power. And procedures designed for "positive" discrimination against such skills could well be considered unfair by those being disadvantaged thereby. But there does remain some possibility of consensus on fairness of procedure in allocating a scarce good, or at least on what would constitute an *increase* of fairness in procedure. We shall use this last idea as the basis for our analysis of fairness in decisions on risks.

6.5 Increasing Fairness in Decisions

Both the "individualistic" and the "public interest" approaches to the idea of fairness of risk envisage essentially simple procedures for decision. In the individualistic version, the risk is to be "offered" to those exposed and they could effectively veto it until their personal costs and benefits balanced satisfactorily. In the "public interest" approach, the controllers would decide, on society's behalf, what is "acceptable", and that would be the imposed risk. But the different interests in a hazard develop a social dynamic, involving values, perceptions and power relations. Once a risk is called into question, a decision must be taken: at what level of severity, if any, should the risk be allowed to persist? All the interests may be involved in this decision, though inevitably in different ways and at unequal strengths. The decision on creation or persistence of a risk may be manifestly unfair, and yet not rejected by those enduring it; thus a risk may be "acceptable" *de facto*, while being "unacceptable" by any ethical criteria (see **5.3.5**). For this reason, we shall henceforth concentrate our attention on the fairness of the decision about the risk, rather than the "acceptability" of the risk itself.

All those involved in a hazard have genuinely different values and different perceptions of the risk; and also the relations of power are generally weighted against those who experience the risks. We can therefore imagine criteria for the fairness of risk-decisions to be based on a redress of this imbalance in decision procedures. This cannot be realistically imagined in terms of requiring complete satisfaction of every person exposed to the risk, nor of enforcing an identity of perception and interest among the sides. But a practical step towards increasing fairness in risk decisions would be to provide facilities whereby those experiencing the risk could be competently advised on it, by someone who is directly answerable to them as a group, and who is given standing by all the other interests in any negotiation.

6.6 The Development of Understanding of Risks

It might be objected that all this ethical analysis provides little guidance for any practical decisions on the level of risk to be permitted in any particular case. Even to arrange for participatory involvement and for risk advice will be a lengthy affair; and when all the interests are competently represented around the table, how is a consensus on an "acceptable" or genuinely fair risk to emerge? We have already indicated, indirectly, that this might be the wrong question to ask. In discussing the ethics of the question, we restricted our position to "increasing fairness of decision", and we have also mentioned the importance of seeing each particular decision in an extended time-scale (**5.3.4**). Hence the dynamics of risk decisions are not a clash between rigid and diverse perspectives. The whole process is an ongoing educational experience for the individuals, groups and communities involved. Our perceptions of risk and our standards of what is "acceptable" are constantly changing, usually in the direction of greater concern for the wellbeing of people and their environment. We are more careful today than we were a decade ago, and we should be even more careful in a decade's time. The process has been well described in connexion with environmental debates, by Professor Laurence Tribe of Harvard University: our values are vague and "such inchoate values are crystallized into distinct preferences or criteria of choice only through the concrete process of seeking means to attain them and gradually discovering what such means entail" (41). And Lord Ashby comments: "In other words, values evolve through the choices made in groping towards them, and it is an essential aspect of freedom that we can choose what we value" (42). In risks, as in environmental questions, we should envisage a "spiral" of progress which "must incorporate procedures for its own evolution" (43). The progress should come about by dialogue and mutual education between those empowered to decide, and those who are directly affected by their decisions. In this process, the scientific study of risks (analysed in Chapters 3 and 4) need not be subjected to the extreme judgments of uncritical acceptance or total rejection. Rather, the scientific assessment of risks will evolve along with the other components of the dialogue, its methods and categories being improved by experience and criticism. In such terms, we can imagine an approach towards fairness in decisions and effectiveness in control of risks.

7. EFFECTIVE CONTROL OF RISKS

In the last chapter we observed a transition of the argument from intellectual discussion of judgments towards recommendations for practical activities. We now continue in the practical vein, and see how controls on risk can either succeed or fail in being effective. The difficulties of control stem from the very logic of judgments on risk: we have seen how it is impossible to form precise assessments of risks and objective evaluations of them. What happens in any attempt at control will then depend strongly on the social dynamic of the hazard: the perceptions, values and powers of those who create, who experience, and who control the risk. It is in these terms that we explore the problems of control, and consider recommendations for its improvement.

7.1 Social Problems of Control

7.1.1 The operation of monitoring is open-ended; "who guards the guardians" has no simple, final answer. But controlling is not simply a matter of standing by and watching. It requires the securing of relevant information, the assessment of hazards, the judgment of whether particular risks fall within agreed limits of acceptability, and also the encouragement of good practice and the punishment of the bad. All this might seem to be mainly technical; but an essential part of the guardian's work is consultation with all those involved in a risk. We shall discuss this in some detail, as it indicates a means to the increase of procedural fairness in the acceptance of risks, that we discussed above (Chapter 6).

We have already spoken of the importance of commitment and morale in the regulation of risk; and an inspector needs to consult his clients if they are to understand and respect his work. Indeed, the idea of "consultation" would seem to require no special support in the British context; for risk-control has traditionally involved consultation to a very high degree. As a nation, we are well aware of the dangers of a bureaucracy creating a world of its own administrative concepts and then wielding power by forcing the affected public to conform. However, in the case of hazards, the consultation process has always been rather one-sided; inspectorates have tended to behave paternalistically, consulting with those who impose risks, while considering those who experience them as passive partners. The case of the Alkali Inspectorate is notorious in this regard; criticisms first made in an independent report by Social Audit (44) were substantially endorsed by the Royal Commission on Environmental Pollution (45).

There has never been any hint of a suspicion that the Alkali Inspectorate has been other than dedicated and impartial in its own way, nor that the Factory Inspectorate has not done its very best. But an inspectorate that does not enjoy the confidence of those who experience risks will not gain information from them. It will then remain ignorant of clues to hidden hazards. Moreover, such an Inspectorate will genuinely tend to see reasonable employers on the one hand, and apathetic or irrationally critical employees on the other. Hence, real consultation with those who experience risks is no less vital a part of the regulatory process than any other.

7.1.2 Other forces tend to impel guardians of risk towards the side of those who create the risks. Although it is widely known that power corrupts, we are as a society less well aware that impotence also corrupts, especially when it is linked to responsibility (46). An inspectorate that cannot enforce its requirements must either confess its impotence or conceal it from view. In the latter, natural course, the inspectorate denies all cases of abuse except for the most flagrant: and in so doing becomes implicated in their continuation. Thus, a weak inspectorate is pushed towards identification with those who create the risks, to the detriment of those who experience them. Though the agents may be honourable men, and quite aware of their dilemma, nonetheless they may be powerless to influence an inherently corrupt situation.

Can such things occur? To a significant extent they may be said to have held for the old Factory Inspectorate. It was at the higher political level that the impotence of the Inspectorate was built into its structure. Understaffing was but a symptom of its problems; more severe was the system that prevented it from having any real sanctions against offenders. In the last resort, an Inspector could take an offender to court. But this would only be before a local magistrate, and the local Inspector himself was required to prepare and argue the case against whatever talent the offenders could command. In any event, financial penalties were derisory, and the local bad publicity not crucial for the management in a national or international firm (47). So any threatened prosecutions were essentially only a bluff, as factory managers were well aware.

Also, the common argument that improvements would be "uneconomic" and could result in the closing of a factory and loss of jobs, can frighten inspectors, trade unions and workers alike. So long as none of the other sides has access to information for testing such warnings, the balance of influence in hazard situations will rest with those who create the risks. Things have changed in many ways under the new Health and Safety legislation. But as these changes are still proceeding, and are also the subject of controversy, we can only recommend that they be closely scrutinized by all concerned with the effective regulation of risks. Also, those concerned with those inspectorates that have so far escaped the critical scrutiny given to Factory and Alkali, might well consider these problems.

7.2 Problems of Science and Scientists

7.2.1 Effective guardians must be well-equipped with scientific knowledge, techniques and methods in the assessment of hazards and risks. But the situation of research scientists in the control of risks is very different from the image purveyed in philosophical, sociological and popular descriptions of science. We are conditioned to think of a scientist as an inquirer who freely chooses which part of the unknown world of nature to discover;

then works up his material till his conclusions are as well-founded as the existing methods make possible; and finally freely publishes his results. This picture is based on the experience of academic scientists engaged in independent, basic research. Although they include most of the eminent and publicly-known scientists, they are only a fortunate minority. Most scientists, including the great majority of those assessing hazards, are workers in an organization.

In regulatory work, the employed scientist will not always have the luxury of choice, problems will usually be chosen by his superiors, on their opinion of the hazards needing investigation. He will carry out their instructions for research, under externally-imposed constraints on time and costs. His results may be used as the basis for policy decisions, regardless of whether in his opinion they are sufficiently strong or relevant for that function. Most important, his results are not his property, to publish as he sees fit; but they belong to the corporate agency that employs him, for use at their discretion. This last restriction is in some ways only reasonable, since the work might involve commercial secrets on the one hand or contain statements which could give erroneous impressions of the problems when read by laymen or journalists. But the employed scientist is then not merely cut off from the "colleague community" of researchers; worse, he may see results that he considers urgent and vital being distorted or suppressed for years. In America this problem has led to formal accusations of bias in favour of business against the Food and Drug Administration being laid by members of its research staff (48). In the U.K. the problems are doubtless less severe, but the situation is exacerbated by the blanket laid on all Civil Servants by the Official Secrets Act.

The position of the academic scientist engaged on monitoring work is a mixture of the two just described. How much freedom he has will depend on the degree of his official involvement in the monitoring operation. In any case, he will need to cope with the constraints, technical and diplomatic, inherent in this work. However, an academic scientist has a particular obligation to inform the public of concealed hazards, when his "employed" colleague could face dismissal or legal penalties for the same act. The ethics of "whistle-blowing" are still rudimentary; on this we commend the Council's earlier report on "Superstar Technologies" (49) and also the discussion "On being, and being held, responsible", in Lowrance's study. (50)

7.2.2 The contribution of scientists is crucial in the setting of standards for permitted limits on hazards. In this work, both academic and "employed" scientists are usually involved. In many cases, a hazard is regulated by the device of an officially-permitted upper limit on a quantitative measurement of its supposed major cause. For example, this might be the amount or concentration of a pollutant. No hazard is known to have a threshold below which there is no harm and above which it suddenly becomes significant. Rather, in any study, there will be some data on the effects of the hazard (frequently sparse and indirect) indicating zones of very high risk and perhaps also zones of very low risk. There will also be some information on the processes causing the hazards, and some estimates of the costs of reduction below the existing level. These latter estimates are likely to be rather speculative in the case of any substantial reduction The maximum permitted measure of the hazard that is adopted will be placed somewhere between the "high" and "low" risk zones. Its particular location may depend on some averaging process, and perhaps on a calculation of "diminishing returns" of further expenditure on risk reduction. For example, there may be a calculation of the level at which some "acceptably" small proportion of the population at risk (say 0.1%) will be seriously affected.
No matter how much refined calculation has gone into the risk assessment, the scientific data are inevitably coarse, and estimates of consequences highly inexact. In the last resort the location of the practical limit of "acceptability" will be by fiat, based on personal judgments of those responsible for the decision. Just how difficult it can be to secure a scientific basis for such a limit, and how imperfect can be the decision procedure, is illustrated by the example of asbestos (see Appendix 5). Although this is now a notorious case, it is unlikely to be a unique example of lax control through permitted limits established by a "scientific" committee.

7.2.3 It is regrettable that scientists and experts who participate in the setting of "permitted" or "tolerated" limits in this way, so often described them as "safe". This confusion is another reason why we have preferred to avoid that term in this discussion. It is unfortunate that the press and the public generally accept such limiting values as the "safe" ones, and are allowed to believe that there is no cause for concern until they are breached. Only occasionally do we see regulations embodying distinctions between degrees of risk, as expressed by "warning levels", "action levels" and "impermissible levels". And the only prudent and fair approach, of considering any actual level of risk as "temporarily tolerable", seems to be coming into practice but slowly.

7.2.4 Even when all the scientists are acting in a competent and principled manner, the imbalance in the hazard situation may be reflected in the composition of a standards committee, the duties of its members as they see them, their conception of the role of "science" in the proceedings, and ultimately in their decisions. Because so many qualitative elements enter into judgments about hazards, it is natural and proper for scientists to tend to advance views favourable to any interests they may represent. But who represents those who endure the risk? As workers or as residents they do not command resources for engaging sympathetic experts. If they are formally represented, eg through a trades union, it will usually be by people who cannot meet the other side's experts as equals in technical debate. In fact, they are often quite unaware that there is some committee deciding on standards that will affect their health and safety for years to come.

7.3 Information

Because risks affect us in so many different ways, it will never be possible to organize their control under one single, tidy Inspectorate. The variety of existing agencies, and the resulting confusion of sources of information is to some extent inevitable. But, there is little sign so far of a change in the old traditions of secrecy among important agencies; and so the citizen will still be left in the dark about what may be affecting his health and property.

As an example of the difficulty of finding information, even in a a popular field, we may quote the example of air pollution, described in *Nature:* (51)

"The Commission talks of the housewife whose washing has been dirtied by a breakdown in pollution control at a nearby plant and who may be "disconcerted and irritated" by a request to leave a message on the District Alkali Inspector's 'phone answering machine (inspectors only have part-time secretarial staff). She might be lucky even to find the right 'phone number. In the London Phone *Directory*, there is nothing under "Air", "Alkali", "Clean Air", "Clean Air Council" (which advises the government on air pollution), "Her Majesty's Alkali and Clean Air Inspectorate", or "Pollution". A very smart housewife might just possibly alight on "Environment, Department of the" where lurking under a sub-sub-heading is "Noise, Clean Air, and Waste". Or she might try her local authority for which, if the 'phone book is not too out of date there could be an entry "Environmental Health Services", but no mention of air pollution. But of course, if it was smoke from a bus, she would have to 'phone London Transport. And what about smoke from a car?"

If people are to have some influence on the control of the risks they endure, they will in the first place need access to information. This is not merely a question of inspectorates making some documents available on request. There is an expertise in knowing what is officially available and how to locate it, as well as in using clues about the existence of information that is officially unavailable. The function of public education might be performed by the existing inspectorates; but their existing workloads and traditions make that unlikely. Hence there is a case for a separate, independent, advisory service, helping the work of education of people in communities about the risks they face.

The tasks of the Safety Representatives envisaged in the Health and Safety at Work Act, will probably consist largely of the provision of information on the hazards of the workplace. But for the great variety of domestic and environmental hazards, there is no machinery for bringing science into the service of those who may be suffering from unfair and ill-controlled risks. It would seem perverse to deny people away from work the channels to information that are now provided for them by law when they are at work. We shall return to this point later (**8.3**).

7.4 Eternal Vigilance

Although every existing hazard is known to have a chance of being realized, the task of control is to keep that chance as small as possible. No serious accident should ever be dismissed as resulting from bad luck or coincidence. A proper analysis of a hazard should lead to monitoring arrangements that prevent such coincidences from occurring. This is not a simple affair, for protection is secured by a variety of means, some of them routine (as standard operating procedures in hazardous situations) and some requiring ingenuity and initiative (as envisaging possible accidents and coping with those that occur, always in unexpected ways). Monitoring must be applied to all these measures, and indeed to monitoring operations themselves.

It is easy to see how complacency and apathy cause accidents.

Without constant vigilance, enabling causes of hazards will not be eliminated, dangerous coincidences will occur and consequences will spread in sequence. The morale of an establishment will be quite crucial for the quality of its hazards control. If everyone is concerned only for his own benefit or convenience, no-one will be willing to invest his time and resources in the prevention of unlikely contingencies. Accidents are therefore one of the costs of bad management, as inevitable as waste of materials and low quality of products in an industrial establishment. Moreover, when hazards are realized in such bad environments, they are much more liable to trigger off further events, in a sequential hazard. A management that has been trying not to know about particular dangers is unlikely to be prepared to cope with the complex situation involving technical, medical, social and political elements created by a disaster. The bungling by all the authorities in the poison-gas incident at Seveso, Italy, in 1976 illustrates this point perfectly.

Effective control of hazards cannot be forced on people; they must want it and must be given the means, technical and organisational, that make it possible. For this information, education and participation are essential. In this way, effective control and fair decisions on risks are very closely related.

8. SOME PRACTICAL IDEAS

We have reviewed the logic of the assessment and evaluation of risks, and the social dynamics of fair decisions and effective controls. This analysis has provided some materials for a deeper study of the problems of hazards. Even at this point we can offer some suggestions for improvements in the very uneven state of risk control in the various fields where it is applied. We can do this quite conveniently by reference to the Appendices to this report, where particular case-studies are given.

8.1 Analysis

The first two Appendices provide a survey of the scope and limits of mathematical studies of hazards. In the case of aircraft (Appendix 1) the circumstances of realized hazards enable detailed and reliable models of the causal network of accidents to be constructed. These could then be used as criteria for a design target for the risk level of future safety systems, with a progressive reduction of risks.

The companion study to this, on radiation hazards, (Appendix 2) shows that much reliable information can be gathered despite the difficulties over data discussed in section 2. In small-scale and well-contained contexts, particularly in the case of medical research and treatment, some sort of rough balance of cost and benefit can be struck, for radiation exposure. Also, when one nation dumps radioactive pollution on another in connexion with its independent weapons development, there is an assault on those citizens, however small. But scientific exactness and moral certainty become weaker as we move to consider the more complex problems of total nuclear fuel cycles. The imponderables (eg. the hazards of "nuclear malevolence" and the possible genetic damage from a large-scale release of radionuclides) eventually outweigh the measurable risks. The decision on proceeding into a "nuclear economy" must then be strongly influenced by very general principles of innovation, perhaps "try it and see" or perhaps "stick to the small and familiar". But it is important to appreciate that even if quantitative, scientific analysis is not enough in such crucial decisions, it is nevertheless very necessary, (and it must be done to the highest standards) for an informed decision to be made at all.

8.2 Discipline

The case of the Flixborough disaster (Appendix 3) reminds us that such realized hazards are less the result of "acts of God" than of omissions of management. We can hope that there has been a change in official thinking on this, since the Inquiry at Flixborough spent such immense effort in choosing between two pipes while merely noting that tons of chemicals had been illegally stored on the site. The checklist of safety practices recently produced by the Health and Safety Executive (52) and the list of good and bad practices provided in Appendix 4, show how demanding risk control can be. It appears that under the new legislation (unlike the old) the quality of management is a legitimate concern of the Factory Inspectors, and is at last recognized as being among the causes of a hazard.

How is industrial hazard consciousness among management to be improved? Certainly, when dialogue begins with the new Safety Representatives a fresh approach can be introduced.

But who will educate these many new officers? Some trade unions, the Health and Safety Executive, and independent agencies are taking up this work. But the scientific and theoretical basis for this is very thin indeed. There is as yet only one university department of Safety and Hygiene (at Aston), and no comprehensive textbook of hazard and risk analysis. When one considers the financial and human costs of the preventable accidents that occur all around us, at the workplace, on the roads, and in the home, this neglect by those who sponsor teaching and research in science and technology is an eloquent testimony to the values that implicitly govern priorities in intellectual effort in our society. (53) In the creation of new syllabuses on the principles of risk control, we cannot do better than to borrow and adapt the wisdom of successful areas. The hazards of ammunition (Appendix 4) present a very pure case of the "major hazard": a low probability of events, but each with an enormous cost. The design philosophy of hazards containment there is so basic that it should be memorized by every person controlling a risk anywhere: "whatever is not impossible (for people to do incorrectly) is inevitable". Unless a hazard can be rigorously shown to be astronomically improbable (as for instance a direct hit by a meteorite), it must be expected to be realized sooner or later.

So there must not only be constant vigilance in reducing the chance of each sort of accident; there must be equal concern for reducing its effects, direct and sequential. After the Titanic disaster, ocean liners not only carried adequate lifeboats, but they were also required to have regular boat drills, to be sure that they would be effective. We may suggest a practical principle: Wherever an installation neglects its disaster drills, hazard control is sure to be slack, and real disaster is, therefore, more likely.

8.3 Involvement

The dreadful story of Asbestos (Appendix 5) shows how no formal machinery of legislation, standards, inspection and participation can be enough by itself. The social dynamics of hazards inevitably involve differences of interests, values and perceptions. If the side with the greatest built-in strength is not checked, either by some other power or by its own conscience, there will result an enfeebled inspectorate and a victimized workforce. To assume that all interests are identical in an industrial hazards situation is to destroy the basis for a free, healthy and effective involvement of those who suffer from otherwise uncontrolled risks. People would thereby be denied the right of self-defence against arbitrary and unfair assaults on their person, conducted for the profit or convenience of others.

Recognizing that there are genuine conflicts of interest does not entail making moral judgments on the different sides or on their representatives. In the terms of the social dynamics of hazards we can see how genuine negotiations can take place over hazards, as in other social situations with conflicting values.

Because hazards are so various, there is no single way towards fair procedures. The people at risk must be able to do something to help themselves, otherwise no outside help will be any use. But they can benefit from many sorts of assistance. In the Asbestos case, the Trade Unions were of little help; though that may now be changing. That case was notable for the successful use of the Courts for exposing a scandal, something far more difficult here than in the USA. People with technical competence, and perhaps inside knowledge of the creation of the risk, can "blow the whistle" and offer technical information and advice. And conscientious inspectors work continuously on behalf of those enduring risks of all sorts.

These different methods might well be supplemented by the new role mentioned in connexion with approaching fairness of risks: the creation of "risk advisors", directly answerable to a group at risk, and given standing by the other interests in risk negotiations. In defining this role, we must keep in mind the opposed dangers of having a service rapidly becoming bureaucratized and paternalistic on the one hand, or becoming unskilled and casual on the other. Most of the work of such advisors would be in the provision of information and in consultation on particular problems. In this they would function much like the new Safety Representatives in workplaces; but in this case there will need to be an even greater preliminary task of establishing the foundations of a service. Although every local authority has Environmental Health Officers, with a strong tradition of monitoring diseases and "nuisances", yet the important fields of domestic accidents, road safety, workplace hazards and fires are outside their remit. And the provision of information to the public on all hazards seems to be no-one's job. For this reason we consider that small pilot schemes, based on particular local communities and related to educational institutions, would be advisable. (Appendix 6). A model for such a service could be the Law Centres, which serve many who would otherwise remain ignorant of their rights in law, and thereby supplement the existing facilities. In spite of the many difficulties of organizing such a risks advisory service, we believe that only in such a way can there develop an involvement of people experiencing risks, which is so necessary for fair decisions and effective control.

9. CONCLUSIONS AND RECOMMENDATIONS

9.1 Conclusions

1. The acceptability of risks cannot be simply derived from a scientific study of quantified probabilities, costs and benefits. The human factor influences the analysis at every point. But fairness in decisions and effectiveness in controls of risks can be approached by the use of scientific methods among others, provided that the diversity of human interests, values and perceptions of risks is always respected.

2. The pheonomena of hazards and risks are particularly difficult to study scientifically by the use of empirical data and theoretical models. Quantitative statements about them must always be very approximate and tentative; an apparent precision in such statements may be quite misleading. Human values are implicit in all assessments of risks and estimates of consequences. Scientists involved in assessing or debating risks are in a situation very different from the traditional one of producing "public knowledge" in academic research. The policy implications of the work impose many constraints (chapters 3, 4 and 7).

3. For risks to be reduced there must be constant vigilance and commitment on all sides. (chapter 7). The principle of human frailty — "what is not impossible (for people to do incorrectly) is inevitable" — must be applied to all arrangements for monitoring and containment. A full awareness by those exposed to risks, and education in risk management, are both necessary for effective control. (chapter 8).

4. The judgment of "acceptability" of a risk involves a consideration of its perceived costs and benefits in the light of feasible alternatives, *by the person exposed to it*. Imposed risks may therefore be "accepted" in practice while being unfair or even intolerable to those enduring them (chapters 5 and 6). We therefore avoid the indiscriminate use of the concept of "acceptability". The term "safe" is also very ambiguous, referring perhaps to a risk that is "acceptable" in some sense, or to one that is believed non-existent. We employ it, and its derivatives, sparingly. (chapter 2).

5. The manner in which risks are imposed shows deep social inequalities: excessive risks tend to be concentrated in the homes, communities and workplaces of manual workers and their families. (chapter 5). Any claims to fairness in the distribution of

risks must be backed by positive action to remove this inequality. Hazards of all sorts have been shamefully neglected in scientific and technological research. (chapter 8).

6. Fair decisions and effective controls require a recognition of the social dynamics of the risk situation. This includes the three interests (those who create, those who control, and those who are exposed to the risk), each with its own values and perceptions. It is unrealistic and detrimental to imagine that these three interests should and do have identical views. (chapter 7).

7. Controls can, in practice, be rendered ineffective by the weakening of the regulatory agencies, to the point where they may become complicit in the unfair imposition of risks. (chapter 7). The involvement of those at risk, through education and participation, is necessary for remedying or preventing this abuse. (chapters 6 and 8).

8. Fairness in risks cannot be achieved by applying abstract principles, either of individual veto or of societal fiat. An approach towards fairness in risks can be defined in terms of the procedures for decisions on risks. This is enhanced by the involvement of those at risk, and provision of information and advice to them. (chapter 6).

9. The problem of risk management should not be seen statically as if it could be solved satisfactorily once and for all by scientific analysis or administrative procedure. Rather, the social dynamics of risks extend to the perceptions and evaluations of the risks themselves. Risk management should evolve through an interaction of all the diverse interests concerned in each case. (chapter 6).

10. The balancing of risks against benefits, especially in the case of a technology not yet in being, is an exercise in which facts are necessarily few and speculations inevitably abound. Any assurances that a particular future technological risk is negligible cannot be based on a scientific proof, but must, in the final analysis, involve a judgment.

9.2 Recommendations

Our single major recommendation is that those who are exposed to risks which are not immediately obvious to them should have a powerful voice — expressed responsibly and on full information and sound advice — in deciding what risks they should be exposed to.

Beyond that, this report can do little more than to sketch the problems of risk management, and to indicate areas where further study would be useful. Our consequential recommendations are for studies (which may be accomplished by research, working parties, conferences, or pilot schemes) that could be successfully undertaken in independent small-scale projects. Among them are:-

1. Conducting exploratory surveys of the status and functions of existing institutions concerned with risk.

2. Examining the research and the standards-setting of particular agencies.

3. Promoting pilot projects on "risk advisors" in particular localities and occupations.

4. Establishing projects for education and debate on risk, using the facilities of the media.

5. Planning for parallel studies of medical risks.

Such studies could be organized through an office, utilizing a small permanent staff of research and information officers, occasional publications and seminars, and periodic conferences for policy recommendations and exchange of ideas, involving representatives of the leading interests in the field. The Council for Science and Society could undertake this work as part of its ongoing programme.

REFERENCES

(1) LOWRANCE, William W.: *Of Acceptable Risk*, William Kaufman Inc., Los Altos, California, 1976.

(2) WEINBERG, Alvin M.: "Science and Trans-Science" *Minerva* 10, 1972, pp. 209-222.

(3) LOWRANCE, W. W.: op. cit. p8.

(4) Ibid. p. 87.

(5) See STARR, C: "Social Benefits versus Technological Risk". *Science*, No. 165, 1969, pp 1232 — 1238. Also STARR, C: "Benefit-cost studies in Sociotechnical Systems" in *Perspectives on Benefit-Risk Decision Making*, Nat. Acad. Eng., Washington DC, 1971.
A more recent study on this subject has recently been published: STARR, C; RUDMAN, R.; WHIPPLE, C.: "Philosophical Basis for Risk Analysis" *Annual Review of Energy* No. 1. Annual Reviews Inc., Palo Alto, California, 1976.

(6) KNOX, E. G.: "Negligible Risks to Health", *Community Health* No. 6, (5), 1975 March-April, pp. 244-251.

(7) Report of the Inquiry into the Collapse of Flats at Ronan Point, Canning Town, HMSO, London, 1968, p42, para 143.

(8) EDDY, Paul, POTTER, Elaine & PAGE, Bruce (of the *Sunday Times* Insight Team): *Destination Disaster*, Hart-Davis, MacGibbon, 1976.

(9) TURNER, Barry A.: "An examination of some of the organisational pre-conditions associated with some major disasters". Presented to the Open University Seminar, City University, London, November 1974. Reprinted in PETERS, G. (ed): *Human Factors and Systems Failures*. TD 342 Unit 4, Milton Keynes: The Open University Press.

(10) Advisory Committee on Major Hazards: First Report. HMSO, London, 1976.

(11) WEINBERG, Alvin M.: "Social Institutions and Nuclear Energy", *Science*, No. 177, 7 July 1972, p.27.
KNEESE, A. V.: "The Faustian Bargain", *Resources* No. 44 September 1973.

(12) WILLRICH, Mason & TAYLOR, Theodore B.: *Nuclear Theft: Risks and Safeguards*, Ballinger, Cambridge, Mass. 1974.

(13) "Reactor Safety Study: An Assessment of Accident Risks in U.S. Commercial Nuclear Power Plants". (Rasmussen report) U.S. Atomic Energy Commission, Rep. No. WASH-1400 (Draft) US-AEC, Washington D.C. 1974.

(14) FARR, C. F.: "Industrial Accident Data: How Much Scope is there for Improvement?", in HENDRY, W. D. (ed), *The Collection, Analysis and Interpretation of Accident Data*, Safety in Mines Research Establishment, Sheffield, 1975.

(15) KLETZ, Trevor A.: "Hazard Analysis — a Quantitative Approach to Safety", The Institution of Chemical Engineers Symposium Series, No. 34. 1971.

(16) SENNECK, C. R.: "Over 3-day Absences and Safety", *Applied Ergonomics*, No.6.3, 1973, pp. 147-153.

(17) LOWRANCE, W. W.: *op. cit.*, p 172.

(18) LOWRANCE, W. W.: *op. cit.*, p 50.

(19) *The Guardian*, 19th March 1976.

(20) TAIT, Nancy: *Asbestos Kills*, The Silbury Fund, London, 1976.

(21) HAFELE, W.: "Hypotheticality and the New Challenges: the Pathfinder Role of Nuclear Energy" *Minerva*, Vol XII, No. 3, July 1974, pp. 303-322.

(22) NORMAN Colin: "Nader embarrasses the US-AEC", *Nature* Vol. 247, February 8, 1974, pp 331-332.

(23) MELINEK, Stanley J.: "A Method of Evaluating Human Life for Economic Purposes", *Accid. Anal. & Prev.* Vol. 6, 1974 pp. 103-114.

(24) STARR, C.: "Social Benefits versus Technological Risk", *Science* No. 165, 1969, pp. 1232-1238.

(25) OWTAY, Harry J.: *Risk Assessment*, the joint IAEA/IIASA Research project, Vienna, 1975.

(26) DAWSON, R. F. F.: "Costs of Road Accidents", *Road Research Laboratory Report* LR 79 (1967), quoted in ADAMS, J. G. U.: "London's Third Airport", *The Geographical Journal*, 1971, Vol. 137, part 4, pp.468-504.

(27) KNOX, E. G.: op. cit.

(28) *The Guardian*, 5th July 1976 p.3.

(29) Commission on the Third London Airport, Report, (Chairman, the Hon. Mr. Justice Roskill) HMSO, London 1971.

(30) DYSON, Freeman J.: "The hidden cost of saying no", *Bulletin of the Atomic Scientists*, June 1975, Vol. 31 (6).

(31) COUNCIL FOR SCIENCE AND SOCIETY: *Superstar Technologies*, Report of a working party. Barry Rose (Publishers) Limited, in association with the Council for Science and Society, 1976.

(32) BLACK, Sir Douglas: "The Use of Research" *Journal of the Royal College of Surgeons in Edinburgh*, Vol 20, pp. 355-364, November 1975.

(33) TUDOR-HART, J.: "Data on Occupational Mortality, 1959-63" *Lancet*, June 22, 1972, p.192.
PRESTON, Barbara: "Statistics of inequality", *Sociological Review*, Vol. 22, No. 1, 1974.

(34) OTWAY, Harry J. & PAHNER, Philip D.: "*Risk Assessment*" *Futures*, April 1976, pp. 122-134.

(35) STEINER, J.: "Risk-Taking", *Proc. Roy. Soc. Med.*, Vol. 63, December 1970, pp. 1271-1273.

(36) WILSON, Richard: "The Costs of Safety", *New Scientist* Vol. 68, No. 969, October 2, 1975, pp. 24-25.

(37) KNOX, E. G.: Op. cit.

(38) FLOOD, Mike E.,: "Nuclear Sabotage", *Bulletin of Atomic Scientists*, Oct. 1976, pp. 29-36.

(39) "Calculating radiation risks", *Nature*, Vol. 260, No. 5547, March 11, 1976, p.100.

(40) RAWLS, John: *A Theory of Justice*, Oxford University Press, 1972 NOZICK, Robert: *Anarchy, State and Utopia*, B. Blackwell, 1974.

(41) TRIBE, L. H.; Schelling, C. S. & Voss, J. (eds): *Where Values Conffict*, Ballinger, Cambridge, Mass., 1976 (To be published in the U.K. by Wiley).

(42) ASHBY, Eric: "Towards an Environment Ethic" *Nature*, Vol 262, No. 5564, July 8 1976, pp. 84-85.

(43) TRIBE, L. H.: Op. Cit.

(44) *Social Audit:* "The Alkali Inspectorate", Report No. 4. 1974.

(45) *Air Pollution control: An integrated approach:* Royal Commission on Environmental Pollution, 5th report, Cmnd. 6371, HMSO, 1976.

(46) RAVETZ, J. R.: *Scientific Knowledge and its Social Problems*, Oxford University Press, 1971, p.422.

(47) GILLIE, Oliver: "The low cost of Risking Factory Workers' Health", *Sunday Times*, May 23 1976, p.3.

(48) GREENBERG, Daniel S. "Review Panel says FDA fails to rebut staff charges", *Science and Government Report*, Vol. V, No. 20, November 15 1975, pp.4-5.

(49) COUNCIL FOR SCIENCE AND SOCIETY: *Superstar Technologies*, Op. cit.

(50) LOWRANCE, W. W.: Op. cit. pp. 119-126.

(51) "Negotiate Flexibly, but Explain Publicly", *Nature* No. 5540, January 22 1976, p.165.

(52) *After Flixborough*, Health and Safety Executive, London, 1975.

(53) CLUTTERBUCK, Charlie, DALTON, Alan & SOLANDT, Andy: "For those in peril: 2" *Nature*, Vol. 252, No. 5481, November 22 1974, pp.262-3.

Appendix 1.

ACCIDENT RATES AND ACCEPTABLE RISK LEVELS FOR AEROPLANES AND THEIR SYSTEMS.

by D. V. Warren (Civil Aviation Authority).

Discussion paper prepared for RTCA Special Committee 130.

1. Accident Rates

A commonly used measure of aviation safety is the *number of fatal accidents per million hours flown.* For world-wide scheduled air transportation (excluding China and the USSR) this has decreased pregressively from a value of about 4.0 in the early 1950's, to 2.3 for 1974.

Another measure which expresses more closely the risk to the individual, and is more easily compared with other means of transportation, is the *number of deaths per hundred million passenger miles,* as follows.

Aviation (1974 — world wide scheduled)	0.38
Pedal cyclists (1974 — UK)	13.3
Motor cycle riders	32.0
Motor car drivers	1.3
Motor car passengers	0.55
Public service vehicle drivers	0.2
Public service vehicle passengers	0.15
Heavy freight vehicles drivers	0.8
Train Passengers (1973 — UK)	0.23

For aviation, the landing and take-off represent the periods of higher risk; and thus the accidental rate is probably more related to the number of flights flown than to distance covered or time airborne, and for 1974 was 2.9 *fatal accidents per million flights.* To express the risk to the individual, the following comparison is made, based on the *number of deaths per million passenger journeys.*

Aviation	1.8 (average journey 475 miles)
Motor car passengers	0.027 (assumed*av. journey 5 miles)
Trains	0.059 (average journey 26 miles)

*This figure was not readily available.

2. Risk to the Traveller.

Yet another way of viewing the 1974 aviation safety index is that it corresponds to 1.43 deaths per million passenger hours, and that is roughly equivalent to the probability of death per hour for the average individual approaching 60 years of age, leading a normal daily life.

If there is a general conclusion from the statistics it is that aviation safety is rather better than the safety of personal vehicles (cars, motor cycles etc) if distance travelled is the criterion, but it is rather worse than the safety of other public services, whatever the yardstick. It is not so much worse, however, that it should significantly affect the risk incurred in a year by an average person.

3. Acceptable Risk Levels for Aviation

In specifying a risk level for design requirements for new aeroplanes, it is argued (Refs 1 and 2) that the aim should be a steady increasing level of safety, for three main reasons:

1) The present position of aviation relative to other means of transport (see paragraph 1 above);
2) The growth of aviation, and the desire to keep the number of accidents occurring in any one year to a reasonable level;
3) The need for the safety of a new aeroplane to match the safety of equipment it is replacing (typically the safety of a type improves as it matures and has the problem areas resolved).

This has led to the acceptance in the UK of a target risk level for engineering or airworthiness causes to be three fatal accidents per ten million flying hours (3×10^{-7} per hour) for future aeroplanes. This represents an improvement by a factor of about three on the existing level for accidents having a basically airworthiness cause, and equivalent improvements are being sought for the operational type of accident.

4. Requirements for Aeroplane Systems.

Engineering accident causes may be grouped under 4 main headings:

(a) Systems
(b) Structures
(c) Powerplant
(d) Flying qualities and performance.

There are many different ways in which the risk level can be apportioned, but if each heading were permitted to contribute an equal amount to the risk of a fatal accident, then 0.75×10^{-7} fatal accidents per hour would be allocated to each.

Turning now to the aeroplane systems, these will range from the flying controls and automatic pilot to cabin environment, communication and navigation. There will be mechanical, hydraulic and electric elements to each. Considering all the elements of all the systems there will be some failure conditions* which are catastrophic in their own right, and many which are not normally catastrophic, but which may lead to a catastrophe in particular circumstances.

Theoretically, a safety assessment procedure could be established for each aeroplane type, which evaluated the probability of each catastrophic failure condition, and each non-catastrophic failure condition (with its hazard level), so that all the possible risks could be summed together to demonstrate compliance with the required safety criterion (0.75×10^{-7}). However, safety assessment is not an accurate science, and many of the systems hazards are not precisely quantifiable, so that some compromise is necessary. This should permit the use of statistical methods where they are appropriate, and engineering assessment and judgment where that is appropriate.

If a risk-summing procedure is to be avoided, then guide-lines have to be established which will ensure that the total risk for the system is nevertheless kept within the desired bounds. The following illustration may assist.

If the total risk of (0.75×10^{-7}) is equally divided between

(a) catastrophic failure conditions — ie those in which catastrophe is certain;

(b) hazardous failure conditions — ie those in which a catastrophe is not certain, but the probability of it is high, say 10%.

(c) major failure conditions — ie those in which a catastrophe is unlikely, say 0.1%,

then there is 0.25×10^{-7} available for each of these categories, *for all systems taken together*.

Considering firstly catastrophic failures, the permissible share of the risk will permit 25 failure conditions, each having a probability of 10^{-9} per hour, but if only two have a probability of 10^{-8}, then only five 10^{-9} failures would take up the remaining risk allocation.

For hazardous failures, 25 failure conditions with a probability of 10^{-8} would meet the target, but any failure conditions with a

---

*A failure condition may be either a single failure, or a combination of failures. For instance, a critical failure may arise if a system fails when the devices designed to protect against that failure have already failed without this being detected.

higher probability consume a disproportionate share of the total cake.

Similarly, for major failures a probability of 10^{-6} is the reasonable target.

In other words, for aeroplane systems:

(1) *Catastrophic failure conditions* which are amenable to statistical analysis should have a probability less than 10^{-9} per hour. Those which cannot be analysed statistically should be so unlikely that they are not regarded as possible.

(2) *Hazardous failure conditions* which are amenable to statistical analysis should have a probability in the range 10^{-7} to 10^{-9} per hour. Those which cannot be analysed statistically should be unlikely to occur in the total operational life of a number of aeroplanes of the type, but nevertheless have to be regarded as being possible.

(3) *Major failure conditions* which are amenable to statistical analysis should have a probability in the range 10^{-5} to 10^{-7} per hour. Those which cannot be analysed statistically should be unlikely to occur to any one aeroplane during its life, but may occur several times when considering the total operational life of a number of aeroplanes of a type.

5. References

(1) LUNDBERG, Bo.: The "Allotment-of-Probability" Shares — APS — Method. International Symposium on Civil Aviation Safety, Stockholm, April 1966.

(2) BLACK, H. C.: *Safety, Reliability and Airworthiness.* International Conference on Structural Safety and Reliability. Washington, April 1969.

Appendix 2.

ACCEPTABLE RISKS RELATED TO NUCLEAR RADIATION

by Prof. E. H. S. Burhop, FRS.

1. Introduction

Problems of acceptability of risk arise in relation to the levels of harmful nuclear radiations to which individuals, groups, or whole populations should be exposed in order to enable the community or sections of it to benefit from applications of nuclear energy.

Although the effects of nuclear radiation are far from being fully understood qualitatively and the magnitude of specific effects on the human organism are not known quantitatively to better than a factor of ten, say, yet the position in this respect is very much better than in most of the other situations we have discussed and it is possible to envisage a meaningful cost-benefit analysis of human activities involving exposure to such radiations.

2. Biological Effects of Radiation.

The biological effects of large doses of radiation are well known. Radiation damages living cells in such a way that when they divide they either die, or produce daughter cells that are non-viable. Tissues composed of dividing cells — eg bone-marrow — are therefore much more sensitive to radiation than tissues composed of cells that do not undergo further division — eg the brain. The germinal tissues of the tests are also very sensitive. The rate of production of blood cells is reduced by the radiation. The white blood cells which are responsible for fighting infection ("the human response mechanism") are first affected, so that in a day or so after a stray exposure, the body loses its capacity to resist infection. The red blood cells are affected later.

Doses of radiation are measured in a unit called the Röntgen Unit (or R Unit). (Sometimes the units rad or rem are used, but for our purpose there is not much difference between them). The dose determines the number of cells damaged per gram of tissue. It says nothing about the total mass of tissue that has received the dose. In dealing with large doses an important concept is the "50 per cent Lethal Dose" (LD50). This is defined as the dose which on average would kill 50% of the exposed individuals. The LD50 appears to be about 400 rem for man. A dose of 800 rem will kill over 95% of exposed individuals. Death will not occur immediately, but will extend over a few weeks. These figures refer to exposures of short duration. The same dose spread over a relatively long period has less severe effects.

There are other longer term but very harmful effects of radiation which of course only become evident when the victim receives a smaller dose, not immediately lethal. These include cancers of various organs, especially of the blood (leukaemia) and of the bone and skin, but also of other organs. These are referred to as somatic effects. Then there are genetic effects due to gene mutations or chromosome aberrations in the germinal cells as a result of exposure of the gonads to radiations. These genetic effects are believed to be invariably harmful. Some "dominant" mutations produce immediately and obviously harmful effects in the progeny of the first generation. Examples are polydactyly (extra fingers and toes), achondraplasia (short-limbed dwarfism), Huntington's chorea (progressive involuntary movements and mental deterioration), and some types of muscular dystrophy, anaemia and eye cancer. Other "recessive" mutations produce effects several, sometimes tens or even hundreds of generations later. Examples are phenylketonuria (a form of mental deficiancy), Tay Sach's disease (blindness and death in the first years of life), sickle cell anaemia, haemophilia, colour blindness and several other types of muscular dystrophy.

These genetic effects range from the almost trivial to the very harmful. Curiously the most lethal mutations are not necessarily the most serious socially. They are so lethal that the foetus may not survive the early stages of pregnancy, so that they give rise to abortive births whose origin is never identified as arising from mutations. Some mutations producing what may seem to be minor defects are responsible for general ill-health. Since their effects are by no means lethal, they may survive in the pool of germ plasm for many generations, and their overall effects may be very serious. These effects lead to a general overall reduction in the average life-span amounting to about 1% for a dose of 100 rem.

There are difficulties in quantifying the risk from radiation. Most evidence of the effect on human populations has been obtained from major industrial malpractices, exposure of radiologists or technicians to radiation, radiation effects of Japanese populations exposed to the atomic bombs on Hiroshima and Nagasaki and of island populations in the Pacific exposed to fall-out from US nuclear weapon tests. These results refer to exposures much greater than those likely to be met with by the population under peace-time conditions. It is necessary to extrapolate from effects of these comparatively large exposures down to the low level exposures relevant under peace-time conditions. How should this extrapolation be carried out? Quantitative evidence has been obtained by studies of the effects of radiation on insects such as drosophila or on mice (the so-called "mega-mouse experiment"). But the response to radiation is quite different in men from that in mice and even more different from that in drosophila.

For genetic effects there is increasing evidence that the number of mutations produced is proportional to the radiation dose down to the smallest doses. This view is consistent with the most acceptable model of the way the genetic effect is produced. For somatic (including carcinogenic) effects the situation is less clear. For some cases it appears likely that the effects may be proportional to the dose. In other cases there seems to be evidence of a threshold. In assessing radiation risks it is usual to make the apparently most conservative assumption of direct proportionality between effect and dose, so that an upper limit of the radiation risk is obtained. For drawing conclusions from cost-benefit analyses, however, it is by no means certain that this dose will lead to a minimum risk decision. For example, take the case of the relative risks from nuclear and fossil-fuel power stations. In normal operations the release of sulphur dioxide into the atmosphere produces damaging effects on vegetation and may lead to deaths from bronchitis and similar ailments. Indeed, it has been estimated that the chemical effluent from coal and oil burning power stations results in about 200 people dying from bronchitis in Britain each year*. According to present evidence, it cannot be conclusively stated that radioactive effluent from the nuclear stations has resulted in even a single death! If, for reasons of being on the safe side, the harmful effects of radiation were grossly over-estimated, this could lead to a wrong and non-optimised energy and safety policy.

3. Doses of Radiation Received by Population and their Quantitative Effects.

Everyone of us is subjected to a natural background dose of nuclear radiation, amounting, on average to about 100 millirem (mrem) per year. About 20% comes from radioactive nuclei, especially in the naturally occuring radioactive isotope, potassium 40, in our own bodies. About 40% of this is due to external gamma radiation from the earth and rocks around us, and especially from the building materials of our houses. The remaining 40% comes from cosmic radiation, entering the earth's atmosphere from outside. The levels of exposure vary a good deal from this average according to where one lives. For example, if one is living in St Ives or Aberdeen where the houses are largely built of granite, the external gamma ray doses may be doubled, while for people living high in the mountains in places like Denver or Quito, the cosmic ray dose may be several times higher.

*Private communication from Professor J. H. Fremlin, Birmingham University.

In addition we are subjected to man-made radiation, from medical and dental diagnostic X-rays, a small fraction of it from occupational exposure, from the fall-out of nuclear weapons tests, and, increasingly, from the implementation of a nuclear power programme. In 1972 the National Academy of Sciences and the National Research Council of the USA published a Report of the Advisory Committee on the Biological Effects of Ionising Radiation (BEIR) on The Effects on Populations of Exposure to Low levels of Ionising Radiation (Hereafter called "Report 1"). This is still perhaps the most authoritative discussion of these questions. It deals of course primarily with the position in the USA, but is readily adaptable to the position in the UK.

Table 1, taken from Report 1, gives the average level of radiation received both "whole-body" and "genetically significant" (gonads) from natural and man-made radiation in the USA. The situation with regard to man-made radiation is very much worse in the USA than in the UK, because of a wider use of X-rays by general practitioners there. This subject was explored by the Adrian Committee, who found that for the UK, the figure for medical and dental radiation is closer to 20 m rem/year than the 73 m rem/year of the Table.

Table 1	m rem/year Whole body exposure	Genetically significant exposure.
Natural Radiation		
Cosmic Radiation	44	
Radionuclides in body	18	
External gamma radiation	40	
	102	90
Man-made radiation		
Medical and Dental	73	
Fall-out	4	
Occupational Exposure	0.8	30-60
Nuclear Power (1970)	0.003	
Nuclear power (2000, projected)	< 1	

As quantitative knowledge of the biological effects of radiation has increased, the allowed levels of exposure have been drastically reduced on several occasions, the most recently being in 1954 when the International Commission on Radiological Protection (ICRP) recommended that radiation doses received by employees by virtue of their occupations should not exceed 15 rem per year.

The US National Committee on Radiation in 1957 reduced this to 15 rem per year. For the whole population they set the limit of 170 millirem per annum for the total radiation dose received, excluding medical radiation. This corresponds to a total exposure of 5 rem before the mean age of reproduction (approximately 30 years). This is related to the amount of radiation estimated as required to produce a doubling of the spontaneous mutation rate, estimated as almost surely between 5 rem and 150 rem and probably between 30 rem and 80 rem.

The BEIR Committee gave quantitative estimates of the effect of a radiation dosage of 5 rem per generation on a population of one million from the point of view of genetically related illnesses and defects.
These figures are given in Table 2.

Table 2.
Number of genetically related illnesses in a population of 1 million exposed to dose 5 rem of radiation per generation.

Disease Classification	*Current Incidence*	*Effect of 5 rem per generation* 1st generation	*Equilibrium*
Dominant Diseases	10,000	50-500	250-2500
Chromosomal and Recessive Diseases	10,000	Relatively slight	Very slow increase
Congenital Anomalies	15,000		
Anomalies expressed later.	10,000	5-500	50-500
Constitutional and Degenerative Diseases.	15,000		
TOTAL	60,000	60-1000	300-7500

They suggest that an additional radiation dose of 5 rem per generation would lead eventually to an increase of from 0.5 to 12.5 per cent of known genetically determined illnesses. Since they thought it reasonable to suppose that about 20 per cent of all ill health is genetically determined, they estimated this level of dosage would lead to an eventual increase of between 0.5 and 5.0 per cent of all illnesses.

Table 3 shows similar estimates for the number of deaths per year due to radiation-induced leukaemia and other forms of cancer for a population of 1 million exposed to radiation of

170 m rem per annum. Estimates of the increased incidence of cancer vary from 1 per cent to 4.5 per cent, depending on the model used to extrapolate from much higher dosage rates.

Table 3.

Number of deaths per year due to radiation-induced cancer for a population of 1 million exposed to radiation of 170 milirem per annum

	Excess deaths due to:	
Age at irradiation	Leukaemia	All other cancers
in utero	0.5-0.65	0.5-0.65
0-9 years	0.8-1.4	0.6-50.0
10+ years	2.4-5.0	9.0-20.0
Total	4.4-6.3	10.0-70.0

4. The Balance of Risk and Benefit in Relation to Nuclear Radiation.

All nuclear radiation carries the risk of harmful biological effects. If there were no benefits, there would be no excuse for allowing any additional exposure to such radiation. Benefits can arise to an individual who may be suffering from a dangerous illness that can be treated by using nuclear radiation. It could be that death is very likely without radiation treatment. Quite large doses are clearly acceptable in such a case. This is an example of a case where the alternative to the use of the radiation involves a risk greater than that of the radiation itself.

Benefit to society may arise from processes which involve an increase in the radiation exposure to operatives or to the general public. The obvious example is nuclear power. One has then to inquire whether the economic and social benefits associated with the extra energy available out-weigh the cost, both economic and in increased human suffering, due to the additional level of illness and death due to the additional dosage of radiation. One has also to inquire once again whether alternative sources of energy, such as the use of fossil fuels or solar power, would involve greater or lesser risk than those of nuclear power.

Despite the difficulty of producing quantitative answers to such questions, even crude estimates may be useful in deciding policy. An attempt can be made in this case, using the figures given in Tables 1-3. From Table 1, we see that the US nuclear power programme envisaged for the year 2000 would imply an additional radiation dose of $>$1m/rem per year per head of population. This figure is based on an estimate of a whole body dose from power reactor effluent of 5 m rem/year at each reactor boundary. The projected level of nuclear power per head of population is considerably greater in the US than in the UK. On the other hand, owing to the greater population density of the UK, the required maximum of 5 m rem/year dose at the reactor boundary will correspond to a greater mean exposure of the whole population. We take therefore the figure of 1 m rem per year per head of population for the UK.

We take the total annual cost of the National Health Service as £7500m for a population of 50 million. We have seen that 5 rem per generation of radiation exposure would produce an increase of from 0.5 to 5.0 per cent in incidence of illness when equilibrium is reached, so that an extra 30 m rem per generation would produce a corresponding increase of between 0.003 and 0.03 per cent, ie between £0.225m and £2.25m in the cost of the National Health Service.

From Table 3 we see that this level of dose in a population of 50 million would produce an additional 50 x (15 — 75) ÷ 170, ie between 5 and 25 deaths per year from all forms of cancer. This is to be compared with present estimates of 200 deaths per annum due to the effluent from conventional hydrocarbon power stations — a figure likely to be much larger due to the increased capacity of such stations expected in the year 2000.

There is little doubt that if this were the whole story the balance would come down decisively in favour of nuclear power. There are however other factors. In the first place, one has to consider the much greater radiation doses that may be received by workers employed in the power station, who may be subjected to a dose of 5 rem per annum, 5,000 times higher than the average received by the general population as a result of the operation of the nuclear power station, so that for each 10,000 employees of the nuclear power station, the total increased medical costs would equal the cost to the whole population estimated above. Further, a worker receiving a dose as high as 5 rem per annum during his working life (30 years, say) would be expected to suffer an average reduction of one year in his life span.

Again no account is taken of the social costs of the mining of the uranium. Uranium miners are at high risk for the development of lung cancer and other radiation-assocated conditions. This has to be balanced against the known dangers associated with coal mining or oil extraction.

Even greater uncertainties hang over the dangers associated with reactor accidents leading to the dispersal of far greater levels of radioactivity and the exposure of portions of the population to large radiation doses. There are dangers associated with sabotage and the theft of nuclear material. These dangers cannot really be convincingly quantified.

Finally, no account has been taken of the dangers associated with the disposal of radioactive waste products. Low grade radioactive waste is currently disposed underground or into rivers or the sea. High grade radioactive waste is concentrated and stored for future generations to deal with. It will remain dangerous and require storage under special conditions for thousands of years.

Clearly, a comprehensive and convincing discussion of the acceptability of risk associated with nuclear power stations is a major exercise and scarcely yet possible. What can be said, however, is that *in normal operation* at least a crude cost-benefit analysis suggests that nuclear power stations are not more hazardous than fossil fuel stations.

Less detailed criteria are commonly applied to assess the acceptability of risk associated with nuclear radiation. For example, the expected eventual contribution of a nuclear programme to the dosage received by the general population was estimated at around 1 m rem per year. This is one per cent of the background radiation. As we have seen already, some human settlements are subject to background radiations twice as large as the average and life appears viable under these conditions. It is unlikely then that an increase of one per cent in the background radiation level is likely to give rise to permanent disquiet.

5. Acceptability of Risk arising from Discharge of Low Level Radioactive Waste into the Sea.

Although a comprehensive discussion of the acceptability of risk in relation to the development of nuclear power as a whole is scarcely possible at present, specific narrower problems associated with nuclear power development are capable of solution. The low level radioactive waste from the Windscale nuclear-fuel recycling plant is discharged into the Irish Sea through two pipes running 1.5km out to sea. A comprehensive estimate of the risk to human population involved in this method of disposal was made before it was accepted as standard practice. The technique of the critical path was used for this purpose. Various paths by which sections of the population could be exposed to radiation from the waste were investigated, and a critical path which led to the greatest exposure was identified. The total exposure of this group was estimated. It was then assumed that if their level of exposure was acceptable, the method of disposal of the low level waste was also acceptable.

Surveys were made of the radioactive contamination of the sea-water itself, sea-weed (porphyra umbilicalis and fucus vesiculoses), fish (plaice), sea-bed mud, shore sand and shore silt in the neighbourhood of the effluent pipes.

The various paths by which radiation could reach man and the relative importance of the overall doses received are shown in Table 4. The critical path obviously comes from eating seaweed. The seaweed porphyra umbilicalis is collected along the Cumberland coast and is used in the manufacture of laver bread in the South Wales area. It was found that the laver bread eaters could be divided into two groups. The larger of these consumed less than 75 grams per day, with an average of 15 grams per day. The smaller group consumed between 75 and 388 grams per day, with an average of 160 grams.

Table 4

Method of Intake or Exposure	*Relative Importance.*
Eating seaweed collected on the shoreline	1
Handling fishing gear used near dispersal point	0.05
Sun-bathing on beach	0.05
Eating fish caught near dispersal point	0.01
Bathing in sea . . . swallowing sea water. . . . children eating sand . . . use of sea-weed as fertiliser.	very small.

This critical group represented only some 100 individuals in the entire laver bread eating population. The hazards that the laver bread eating population are exposed to are somatic rather than genetic and caused by the internal irradiation of the skeleton, the gastro-intestinal tract and lower large intestine. Leukaemia and bone cancer are the two most common maladies to be expected. The critical group of laver bread consumers was estimated to receive a dose of 0.4-0.7 rems per year to the gastro-intestinal tract compared with the maximum recommended dose by the ICRP for the general public of 1.5 rems per year. The single largest consumer, with a voracious appetite for the laver bread, consuming 388 grams per day was estimated to receive a dose of 1.3 rems per year.

The genetic hazard was negligible since only a very small section of the British population (about 50,000) eat laver bread at any time. The total radiation to the whole British population, which determines the genetic hazard, is therefore extremely small. It was concluded, therefore, that the risk associated with low level discharge into the Irish Sea was acceptable. (More recently the collection of porphyra from Windscale was discontinued).

6. Acceptability of the Risk to the Population of Australia from Fall-out from French Nuclear Weapons Test Explosions in the Pacific Ocean.

In 1973 the Government of Australia was concerned at the possible effect of the fall-out over that country following French nuclear weapons tests in the South Pacific. The Report of the United Nations Scientific Committee on the Effects of Atomic Radiation to the General Assembly, 1972 Table 45 (vol. 1, p.95) (hereafter called "Report 2") gives the dose commitment in the Northern and Southern hemisphere temperate zones for the total exposure to the gonads, bone-lining cells and bone marrow as a result of all tests carried out before 1971. This Table is reproduced here (Table 5). In brackets are given the corresponding figures for the tests carried out before 1969.

Table 5.

Dose Commitments (m rem)

North Temperate Zone			South Temperate Zone		
Gonads	Bone lining cells	Bone Marrow	Gonads	Bone lining cells	Bone Marrow
170	260	230	55	81	72
(110)	(240)	(170)	(33)	(64)	(47)

The differences between the two sets of figures have to be attributed mainly to the French and Chinese tests, the former accounting predominantly for the Southern Hemisphere increase, the latter for the Northern Hemisphere increase. The increase over the three year period is seen to be proportionately greater in the Southern than in the Northern Hemisphere. Recalling that two high-yield (megaton) French tests were carried out already in July-September 1968, it seems reasonable to assign half the total dose commitment of the pre-1971 tests to the French tests in the South Pacific, ie dose commitment to the gonads of 27 m rem, to the bone-lining cells of 40 m rem, and to the bone-marrow of 36 m rem. Using estimates of genetic effects given in

Report 1 (p.54) a gonad dose of 27 m rem will produce the numbers of ill-effects per million live births given in Table 6.

Table 6.
Number of genetic ill-effects per million live births

	From gonad exposure to fall-out due to French South Pacific tests.	From natural radiation
(a) *Specific genetic damage*	1-10	10,000
(b) *Cytogenetic effects*	1	
i) congenital anomalies	∽1	5,000
ii) abortive births	∽3	55,000

Supposing these effects are apread over 2½ generations for the specific genetic damage and one generation for the other defects, and taking the population of Australia as 12,000,000 with a birth rate of 20 per 1,000, we come up with the following totals:-

Specific genetic damage	18-1800
Congenital anomalies	∽8
Abortive births	∽24

In addition to the abortive births referred to in the Table, there is a much larger class of genetic damage that results in abortion at a stage too early to be detected.

Report 1 considers that the figures in Table 6, calculated for well-defined diseases, are only the tip of the iceberg. "The most tangible measure of total genetic damage is probably poor physical and mental health. Although we cannot measure the personal stress this causes, we can measure morbidity in economic units, such as days lost from work or medical expenses".

The somatic effects include leukaemia, arising especially from strontium 90 ingested in the bone, thyroidal cancer arising from iodine 131 taken up in the thyroid, and many other types of cancer. Young children are especially at risk. At greatest risk are children in the age-group 0-10 whose mothers have been exposed to radiation in the pre-natal period, since the young foetus is many times more sensitive to radiation damage than the child after birth.

From figures given in Report 1 and the estimate given in Report 2 (vol. 1 Table 44) for the total external dose commitment of 35 m rem from fall-out in the temperate zone of the Southern Hemisphere, we expect a total of between 40 and 130 extra deaths from all forms of cancer as a consequence of the French tests.

These numbers are very small compared with genetic and somatic effects produced in the Australian population as a result of natural radiation (see example, Table 6). It might be concluded that they are so small as to be negligible. This would certainly be the case if there were any resultant social benefits. But what social benefit can possibly accrue to the Australian people from the French tests? In such circumstances no increase at all in the risk of any disabling effects among the Australian people can be justified and the Australian Government was plainly right to object to the continuation of the tests.

Appendix 3.

SOME COMMENTS ON THE REPORT OF THE COURT OF INQUIRY INTO THE FLIXBOROUGH DISASTER (1)

by Brigadier R. L. Allen, CBE.

The Terms of Reference given to the Court by the Secretary of State for Employment directed a formal investigation to be held under Section 84 of the Factories Act 1961 into the accident which occurred on June 1 1974 at the Factory at Nypro (UK) Ltd at Flixborough.

The Court decided to limit its studies to those matters which would enable it to establish the causes and circumstances of the disaster as speedily as possible, and to point out only the immediate lessons arising therefrom.

The Court justified the narrow limits of its investigation in para. 8 of their Report. Major factors which led the Court to this decision were:

a) The need to produce a report in short time.
b) The wider implications of hazardous installations would be covered by another Committee (the Major Hazards Committee) whose setting up had been announced by the Secretary of State on June 27 1974.

The Court therefore took the view that they should not "investigate or seek to make recommendations upon such general matters as the proper policy with regard to safety, siting, layout and construction of such plants as that at Flixborough." (para. 8). The Court's arguments for this exclusion were that to have done so "would have involved the taking of an immense amount of evidence concerning the practice at similar plants throughout the world . . ."

This implies that they saw nothing of value in exploring the field between the narrow immediate technical causes of the

Flixborough disaster and the whole world scene of hazardous industrial plants. It appears not to have occurred to the Court that it had a unique opportunity to study in depth all the events in the area covered by the perimeter fence of Flixborough together with the area beyond the perimeter fence where damage to property took place. It is evident that an extension of the Inquiry in this way would have provided information of great value to the Major Hazards Committee, at little extra cost. A unique opportunity existed to gather appropriate witnesses and indeed to widen the questions put to the witnesses who were in fact called. There is no doubt that the Major Hazards Committee will have to study in depth the effect of plant explosions, and if Flixborough is to feature in their investigations, some of the witnesses called before the Flixborough Court may have to be recalled by the Committee at some much later date.

Appendix 4 of our Report deals with the approach to risk management of explosives in Ministry of Defence establishments. Through the work of the Explosive Storage and Transport Committee and by trials and experience there already exists within Government detailed knowledge of the effects of the propagation of a known quantity of explosive and inflammable materials upon dwellings etc located a given distance away, and also the protection required by methods of construction and separation to prevent primary explosions of fires from initiating secondary ones in neighbouring explosive or inflammable materials.

These considerations were not examined or discussed by the Court. There is nothing in the Report to indicate the quantities of cyclohexane contained in each of the reaction vessels. Subsequent inquiries have led me to believe that this information was suppressed for reasons of commercial secrecy. If this is the case, I question such a decision, as the magnitude of the disaster must be related to quantities of material involved.

The only quantitative figures given related to storage. viz:-

Cyclohexane	330,000 gallons
Naphtha	11,000 gallons
Toluene	26,400 gallons
Gasolene	450 gallons
Total Fluids	367,850 gallons

The site was licensed under the Petroleum (Consolidation) Act 1928 to hold only 7,000 gallons of naphtha and 1,500 gallons of gasolene. At the end of paragraph 194(c) the Court states that "the unlicensed storage of large quantities of fluids had no effect upon the disaster." This statement is puzzling, and is capable of different interpretations, but one looks in vain through the proceedings to find the correct one. It could mean that had a fresh licence been sought from the local authority for these large quantities, it would have been granted. Alternatively, it could mean that these quantities of liquids were unaffected by the initial explosion and subsequent explosions and fires, made no contribution to them, and were recovered intact after the disaster.

Following inquiries I made of the Department of Employment I am advised that the first of these two meanings was what the Court had in mind. After the original licence for 8,500 gallons had been granted by the local authority, Nypro brought its fluid storage arrangements into line with new regulations, and therefore there would have been no reason for the local authority to have refused a licence for the quantities actually stored, had the request been submitted by the Company and processed by the new local authority following local government reorganization. It was in fact not the case that these large quantities of inflammable fluids "had no effect upon the disaster" in any literal sense, and they were not recovered intact afterwards. Whilst they had no part to play in the initiation of the first explosion, they nevertheless made a major contribution to heat, flame and smoke when they ignited, and they hindered rescue operations. Indeed, fires were still preventing the recovery of bodies 10 days after the explosion.

I consider this woolly statement by the Court to be a by-product of their narrow obsession with the immediate causes of the explosion, and their relative indifference to the totality of the effects produced by it, which approach I have criticized earlier. Section 84 of the Factories Act 1961 which governed this Inquiry has now been overtaken by s.14 of the Health and Safety at Work Act 1974, and the Health and Safety Commission may now broaden the scope of inquiries as they see fit. I am advised there was some discussion about the scope of the inquiry among members of the Court, but I believe they reached the wrong conclusion in narrowing it as much as they did, and their justification for so doing is unconvincing.

The quantities of fluids held on the site, 367,850 gallons as opposed to the 8,500 gallons for which they were licensed, produced an excess by a factor of 43 over that authorized. All that the Court has to say of this is:- "We recommend review of existing (licensed storage) regulations" (para 233).

"At no point in the Inquiry was there any evidence that the chemical industry, or Nypro in particular, was not conscious of its responsibilities relative to safety. On the contrary, there were indications that conscious and positive steps were continually taken with this objective in mind" (para 201).

"Nypro were safety conscious" (para 202).

"We repeat that there was no evidence whatsoever that Nypro placed production before safety" (para 206).

No reference to the dangers of storing or handling large quantities of potentially explosive or inflammable materials are mentioned under the headings "Lessons to be learned"; (paras 195-208); "Specific Lessons" (paras 209-210); Miscellaneous Lessons (paras 215, 216).

Questions affecting plant layout and construction; siting of plant and licensed storage are covered by the headings "Matters to be referred to the Special Committee or Other Bodies" (paras 217-224).

It should be borne in mind that any serving officer or civil servant who allowed the quantities of explosives in a depot to exceed the limits imposed by Ministry of Defence Regulations by a factor of two, let alone a factor of 43, would be subjected to severe disciplinary action. He would hardly be regarded as safety conscious.

It is striking that the Court of Inquiry did not seek evidence on these matters from the Explosives Storage and Transport Committee. Indeed, in seeking to clarify various issues raised in the Report, I was told by a civil servant that members of the ESTC had been instructed to make no comment to outside bodies upon the Flixborough disaster. The ESTC had expressed interest and had offered to help the Court. These offers were not taken up (paras 217-220.)

The matter of the quantities of cyclohexane being processed as opposed to being stored is a curious omission from the Report, and if there is to be any reform in licencing arrangements either for potentially explosive fluids or for lethal chemicals which can arise even as by-products of a reaction chain, as in the recent accident with dioxin in Italy, this is surely something which must not again be neglected, nor indeed allowed to be concealed by commercial secrecy.

A General Philosophy of Risk Management where Large Scale Explosions can Arise.

The philosophy which is in operation in the Armed Forces in dealing with ammunition, explosive and hazardous materials is:-
First — to ensure that the practices in operation and the skill and training of personnel are at all times such as to ensure that an accidental explosion is an event of low probability.
Second — to accept the inevitable fact that in spite of all precautions it is nevertheless impossible to ensure that the probability of an accident is zero. Hence the effects of an accidental explosion must be anticipated at all points of storage or processing, and positive steps must be taken and be in force at all times to limit the consequences in terms of loss of life and damage to property.

Neither of these measures is effective without the other, and neither is easy or cheap to achieve. To achieve them requires *inter alia* highly trained personnel; standard operation procedures and regulations meticulously drafted, properly disseminated and maintained up to date; immediate reporting and thorough investigative arrangements of all explosive accidents wherever they occur; a dedicated inspectorate with a technical chain of control which can exercise powers of veto without delay; an effective separation of responsibilities for production and for safety, so that production is never won by disregard for safety.

The Court's examination of matters in these areas was cursory. "There was . . . . a Safety and Training Manager, Mr. E. Brenner, whose precise position in the management structure appeared to be somewhat uncertain, but who *regarded himself* as responsible to the Personnel Manager albeit that he had a right of access directly to the Managing Director." (para 23) (my italics).

"A co-ordinating function was exercised by Mr Boynton . . . He was in our view not qualified to act as a coordinator of the Engineering Department of a plant such as Flixborough and should not have been asked to assume the responsibility even for a short while." (para 24)

"Following the departure of Mr Riggall, the Works Engineer (para 24) various responsibilities fell upon others: "None of these were professionally qualified mechanical engineers." (para 26)

From then on, there was no mechanical engineer on site of sufficient qualification, status or authority to deal with complex or novel engineering problems and insist on necessary measures being taken." (para 27)

To anyone experienced in the explosives field, these areas of omission and commission make the Court's praise of Nypro's safety consciousness, quoted above, incomprehensible. Consider the following: "We entirely absolve all persons from any suggestion that their desire to resume production caused them *knowingly* to embark on a hazardous cause in disregard for safety of those operating the plant". (para 57) (my italics). This paragraph relates to the action following the discovery on March 27 that cyclohexane was leaking from Reactor No. 5.

The implications of the word "knowingly" are substantial. The explosion led to the deaths of 28 people and injuries to 36 people on site. "If the explosion had occurred on an ordinary working day, many more people would have been on site and the number of casualties would have been much greater". (para 1). Hundreds of people suffered relatively minor injuries; 53 people were treated as casualties; 1821 houses and 167 shops and factories were damaged.

It is evident that no engineering modification however trivial to a complex and potentially hazardous chemical plant should have been authorized without a detailed technical specification being drawn up by qualified people, when such great consequences might conceivably flow from it. Indeed, for any responsible manager or engineer to deserve to be called "safety conscious" requires that such potential consequences must always be in the forefront of his mind. The word "knowingly" in para 57 is therefore a most serious indictment which the Court appears to have overlooked. It is an indictment of the philosophy operating in the plant; of the chain of command; of the methods of training; of the management hierarchy; and of the operating procedures and regulations. The only reference in the Report which can be found on operating procedures is that in para 194(b) under the heading "Miscellaneous", where the Court says: "There were undoubtedly certain shortcomings in the day to day operations of safety procedures, but none had the least bearing on the disaster or its consequences and we do not take time with them."

Here I must wholly part company with the Court as in my view it is a total abdication of responsibility to run a complex and hazardous chemical plant without the most exact and detailed standard operating procedures, and the disciplines needed to ensure that they are complied with. *Suitable operating procedures in existence at Flixborough would undoubtedly have prevented this particular accident from occurring.* The Court's views are in striking contrast to NASA's, where the most exact procedures exist and are constantly monitored to protect the lives of two or three astronauts.

Para 53 also gives an insight into the philosophy pertaining at Flixborough. When on March 27 cyclohexane was discovered leaking from Reactor No. 5, this must have been recognized as a serious event, calling for immediate action. What took place, however, was that "the Chief Superintendent on duty telephoned the Plant Manager for Area No. 2 and it was agreed between them that the plant be shut down." This must mean that the Chief Superintendent on duty had no authority to shut down the plant, *presumably in any circumstances as those pertaining were extreme*, without reference to a superior.

Further implications concerning operating and safety procedures arise from the following: Pressure testing requirements are set out in BS 3351 para 7.4, and should be by water. "Hydraulic testing of the 20 inch pipe and bellows assembly was never considered." (para 73a)

"'Piping tested hydrostatically shall be tested to a pressure of not less than 1.3 times the design pressure adjusted to 50°C but in no case less than 7 bar'. This kind of test does not seem to have been considered . . . such a test would almost certainly have caused failure of the pipe and bellows assembly and the disaster would have been averted." (para 73c)

After the section was again on stream and the assembly lagged "it was never closely inspected but was casually looked at on frequent occasions by a number of witnesses. One of the witnesses observed that under pressure the pipe seemed to lift slightly off the support pipes, but no-one noticed anything amiss. It must therefore be taken that albeit there may have been some displacement of the assembly during the period, it cannot have been great enough to attract attention." (para 74)

Engineering practices seem to have moved a long way from gauges and micrometers if physical movements must be such as to be readily visible to the unaided eye. Moreover, leaks were allowed to "cure themselves". (para 78).

On Probability.

Paragraphs 191-193 of the Report deal in general terms with probabilities. This is introduced to compare the feasibilities of the two major hypotheses examined by the Court, namely the failure of the 20 inch pipe or the 8 inch pipe referred to as the 20-inch and 8-inch hypotheses. No attempt is made to quantify probabilities, and indeed to have done so would necessarily have involved speculation which the Court rightly rejected. On the 20-inch hypothesis they refer to "a single event of low probability" and on the 8-inch hypothesis "a succession of events most of which are improbable".

The concept of improbability plays an important part in considerations of safety of methods, plants and equipments etc. Nevertheless it is often misunderstood even by well qualified scientists and engineers. Aircraft designers aim for a failure rate of not more than 1 in 10 million operations for various components such as undercarriages. Nevertheless, quite apart from human error, various sequential events each of low intrinsic probability do regularly occur leading to loss of life in air crashes. Every bridge hand dealt has an extraordinarily low intrinsic probability. Yet one has to do no more than deal 52 cards to be confronted with one of them. At one time in the Army it was estimated that the chances of encountering a "rogue" detonator (used for initiating demolition explosives) that is to say a detonator which could go off even with gentle handling, was one in a million. This gave cold comfort to the RAOC Ammunition Inspectorate confronted with a task of that time of inspecting all of them in one depot which had a stock of 10 million. The chance of any given atom disintegrating by radioactive decay is extremely small, but when the population of atoms is large, even on luminous watch dials, the product of the probability and the population of atoms is sufficiently great to ensure virtually continuous radiation.

". . . even the greatest improbability always remains a probability, however small, and that consequently even the most improbable process — ie those which we propose to neglect — will some day happen." Karl R. Popper — "The Logic of Scientific Discovery."

It is of course tempting and plausible to say, when some such improbable event takes place because it must, that the event is on no account to be assigned to its true cause (namely its probability) by reason of the fact that it is highly improbable!

This does not of course make such an argument valid. Nor is it safe in the absence of evidence of causation to assign the choice between two alternatives on the grounds that one is less improbable than the other, especially when the consequence is a disaster which may at some point be repeated because the correct lessons were not learned.

I do not say this to assert that the Court was wrong in assigning the cause to the 20-inch hypothesis rather than the 8-inch hypothesis, because in this case the other evidence is strong, and the Court's introduction of probabilities adds nothing of value to the argument. It is offered because it indicates the caution which is needed in dealing with probabilistic arguments in investigating disasters.

Hence it follows that in considering complex plants and equipments wherein lie possibilities of a major disaster, it is a mistake to draw too much comfort from the fact that any single kind of failure carries a low intrinsic probability. There is in fact always a large population of hazardous contingencies which should be borne in mind. Hence a product of probabilities and events may bring the overall probability of hazard to levels which can be of concern. It is considerations of this kind which lead the Armed Forces to their philosophy of assuming that accidents are inevitable and taking steps to ensure that the consequences are limited.

The Court did not examine the effects of siting reactor vessels at greater distances from one another, or of giving each of them a higher degree of protection by traversing with earth works or brick walls. Although such steps cost money, there is little doubt that had they been in force at Flixborough, much of the site might have survived the explosion, and that the casualties and damage to houses, factories and shops would have been far less. Although the Court sets out in para 8 the reasons why they did not consider such issues, it is felt that by doing so they may have neglected the most important considerations of all. It is certainly something which the Health and Safety Commission, upon whom the Court relied for follow-up investigations, must on no account neglect.

Reference

(1) The Flixborough Disaster — report of the Court of Inquiry, Department of Employment, London, HMSO, 1975.

Appendix 4.

AMMUNITION AND EXPLOSIVES — AN APPROACH TO RISK MANAGEMENT IN A HAZARDOUS FIELD

by Brigadier R. L. Allen, CBE.

The field of risk management covers so wide a spectrum that it is not feasible to produce in a single document a prescription or a code of practice for every contingency — drugs, rail and road accidents, industrial hazards, accidents in the home, deep-sea diving, coal mines, fire services, building sites etc etc.

It may be of advantage nevertheless to examine a single area of substantial hazards to see the kind of things which have to be taken account of — an area moreover with a tradition going back over very many years, and one in which there has been a continuity of experience and expertise. Space necessarily makes this examination less than wholly comprehensive.

The field chosen is that of ammunition and explosives, which are intrinsically dangerous commodities. They are specially designed to kill and maim — but, of course, only on the battlefield. They must not explode during manufacture, movement by road, rail, sea and air, or during inspection and storage under a wide variety of climatic conditions. Explosives and some kinds of ammunition have the property that if one item in a box should detonate, it could propagate a mass explosion and fire causing great destruction over a wide area. A shell in a gun has to withstand enormous accelerations, several thousand times that of gravity, together with very high rates of spin. An internal component of a fuze weighing only an ounce or so under gravity can weigh a ton or more in traversing a gun barrel. A shell must not, of course, detonate in the barrel of the gun even under such conditions, nor prematurely in flight until it reaches its target. Many explosives are chemically unstable, and undergo slow deterioration from the moment of manufacture. To meet the opposing requirements of high lethality on the one hand, and intrinsic safety on the other obviously calls for great design skill and continuity of experience, lengthy trials, meticulous inspection of components both during manufacture and in service, and demanding storage, packaging and identifying criteria.

It is the "user arms" of a Service, say for example the Royal Artillery, or the Infantry, which state the general performance criteria they seek in the development of some new weapon system. This is followed by a feasibility study by design experts to establish whether such criteria can be met within existing or foreseeable technology, or whether some compromise must be sought. In due course, design criteria are finalized between the parties concerned and prototypes are manufactured. These are subjected to careful and meticulous design, handling, climatic and performance trials *by a qualified inter-Service Board totally independent of design, manufacturing or user interests or pressures.* The independence of this Board is prized and regarded as crucial. The Board establishes whether or not the prototype has met the design criteria, the user performance criteria, and all the known criteria of safety in use, in storage and in transport with due regard to the anticipated life in service of the weapon system. The unexpurgated findings of this independent Board are published and distributed to all the many parties concerned.

Serious technical and administration problems can arise and indeed fatal accidents can be caused if the identification of ammunition, components, date and place of manufacture and filling, lots, batches and packages etc., is defective. There are published inter service criteria for markings and for cataloguing of ammunition, components and packages, and this work is covered by appropriate committees. The application of these criteria is verified by inspectors not only at the place of manufacture, but also when the ammunition reaches a Depot. Experience has taught that this double inspection, although it may at first sight look like duplication of effort, is nevertheless essential. There could be damage in transit, and moreover manufacturers work to drawings, and if the drawings are in error as regards final markings, the end product will not conform to the published criteria, and both drawings and early manufacture may have to be corrected by feedback of information about such errors to their source. For all the care taken, errors of this kind are in practice not uncommon. There is also a Committee, the Explosives Storage and Transport Committee, whose function it is to classify ammunition according to its explosives, fire, toxic etc characteristics, and to determine and publish storage criteria. Ammunition is divided into "Groups" which determine inter alia what natures may or may not be stored together in the same storehouse, magazine, or hold of a ship. It would be folly for example, to store ammunition containing white phosphorus (which is spontaneously inflammable in contact with the air if a shell should leak) in the same storehouses as detonators, gunpowder, gelignite or bulk gun propellants. Indeed for shipping purposes white phosphorus ammunition is always treated as deck cargo.

Ammunition is also classified into "Categories". The "category" of a nature of ammunition is usually determined by means of trials which establish whether there is or is not predominantly a mass explosive risk, or fire risk etc.

Buildings in which explosives are stored are marked by signs which give firefighters information about the types of risk they face from the contents in the event of fire or explosion.

An explosion may have effects far beyond the storehouse in which the explosives are stored, and the distance over which these effects are likely to be manifested is catered for by the concepts of "inside" and "outside" safety distances. The "inside" safety distance governs the permitted distance between explosives storehouses in relation to the type and category of the ammunition stored therein. It is such that any explosion in one storehouse will not propagate to explosives in adjacent storehouses. The "inside safety distance" is thus a function of three things:-

(a) the explosive content of storehouses expressed in terms of the actual mass of high explosives (or the equivalent mass of gun propellants, low explosives, etc., derived by factorization).

(b) the type of construction of the storehouses themselves.

(c) whether or not the storehouses are "traversed", ie protected by earthworks. Traverses allow a substantial reduction in inside safety distances, and economise in space.

The "outside safety distance" is that between any given explosives storehouse and inhabited dwellings or offices etc. If a depot exists, then no dwellings, offices, schools etc may be built within this distance, which is calculated to be such that, in the event of an explosion, the damage arising to property will be limited to the superficial. Of course, if a new storehouse were contemplated in an existing depot, and dwellings etc also already existed, then the quantity of explosives permitted to be stored in the storehouse concerned would be circumscribed by the "outside safety distance" available.

The outside safety distance may, and usually does, extend well beyond the perimeter fence of a depot and into property not owned by the Crown. Originally, "yellow lines" on Ordnance Survey maps indicated the outside safety distances around a depot, and this information is available to Local Authorities involved with planning applications. The area covered by the yellow line was based on houses etc constructed with not less than 9 inch brick walls. When architects began to design offices etc with very extensive use of glass, the yellow line distances were no longer adequate for protection, and a purple line was drawn around depots beyond the yellow line to circumscribe the area within which this type of construction would be forbidden.

Taken together, the inside and outside safety distances effectively ensure that the effects of any explosion are constrained. Both these types of "distance" have been established on the basis of considerable research and careful trials, including mass explosions conducted in Heligoland after the Second World War. (It is incidentally remarkable that these well established criteria have not yet been applied to industries where potentially explosive materials are stored or processed in inhabited areas. There is no doubt whatsoever that had they been adopted when the Flixborough plant was designed and built, the consequences of the explosion on June 1 1974 would have been very much less severe than they were, both within the factory, and outside it. The Court of Inquiry made no adequate reference to such considerations).

Three "limits" have been identified so far, namely explosives limits for storage or processing within buildings, and the two kinds of distance limits. There is also another kind of limit, namely the "man limit" (doubtless now called the personnel limit). Any process of inspection or production of ammunition or explosives can always be done, whatever methods are adopted, within the constraint of a labour and supervisory force of a given number of people. Once that minimum has been established as not only all that is necessary but also prudent for the discharge of a given task, there can then be no necessity for that number of people ever to be exceeded. If it is exceeded, it means simply that some people are needlessly exposing themselves to a dangerous environment. In ordnance depots, repair factories, and Royal Ordnance Factories, exceeding the determined maximum number of people is forbidden. Man limits are therefore set in relation to any given task for every discrete working location in a factory; both the explosives limits and man limits are prominently displayed there, and to exceed these limits is a disciplinary offence. Taken together, all these four kinds of limits have the effect that in the event of an accidental fire or explosion, the maximum damage and casualties that can be incurred is, so far as possible, predetermined, and not left to whim or to chance.

The application of limits of this kind illustrates an important philosophy which governs the thinking of people responsible for administration and technical management in a field recognised as hazardous, which can be summed up as follows:-

"*Always assume that in spite of every precaution the maximum credible accident will sooner or later occur, and ensure without fail that when it does, the consequences will be minimal, both in regard to casualties and damage to property.*" This philosphy has withstood the test of time.

There are many kinds of ammunition, explosives and guided weapons, each with its own design problems and idiosyncracies. It should therefore be obvious that personnel responsible for all the tasks called for — the detailed make-up of items; the history of the problems which have arisen with them; the methods of sampling, inspection, proof, modification, repair and disposal; their technical characteristics; the methods of use, storage and transport, and so on — demands considerable training and experience both theoretical and practical. In the Army the task falls to the ammunition technical officers and ammunition technicians of the Royal Army Ordnance Corps who undergo careful selection, followed by the most extensive education and training. (These people are more familiar perhaps to the general public in their role as the officers and men who also deal with terrorist bombs and devices, previously in such places as Aden, Cyprus and Hong Kong, and now in Northern Ireland.) At their head is the Chief Inspector of Land Service Ammunition. He determines training policy, and controls a headquarters which publishes up to date information about all natures of ammunition explosives and guided weapons, and it is to his headquarters that all reports come on accidents, defects, malfunctions, the results of inspection, proof etc. Appropriate information is fed back constantly to users and designers, and to the separate Inspectorate which controls manufacture in Royal Ordnance Factories and by contractors. All ammunition technical officers have direct access to the Chief Inspector and this technical chain of control cannot be interfered with by any intermediary administration or Command hierarchy, nor even by a GOC of Command or a C in C. Hence nothing in the way of expediency can ever override or impair this technical network. Should the use of a particular nature or type of ammunition be banned by the Chief Inspector in the light of the facts, this ban can then be removed only by him following such investigation and advice as is necessary.

The editorial quality of books, pamphlets and technical instructions in an area of technical complexity and hazard is something which needs particular emphasis. Carelessness and ambiguity in drafting, delay in dissemination of information or failure to amend, is something which can cost lives. Nothing less than excellence in all these matters will serve. The attitude to safety must be swift, uncompromising and authoritative. There is not the slightest doubt that administrative failures can lead to accidents and deaths just as inevitably as can design or production failure, for it is the administrative process that must react to them. The following list summarizes some of the administrative and managerial failures which must be constantly guarded against and rooted out, and at one time or another neglect of each of them has led to serious trouble.

Administration

A defective management hierarchy

Separation of responsibility from authority, and inadequate delegation arrangements.

Failure to separate the inspection function from the production function.

Inadequate operating procedures and standing orders.

Poor technical libraries and publications, and a failure to keep them properly amended.

Education and recruitment below the necessary standards.

Defective cataloguing and marking of equipment stores and spares.

Poor communication.

Poor personnel management and an inadequately disciplined and motivated management and workforce.

Poor inspection arrangements and inadequate powers of inspectorates.

Inadequate establishments.

Inability to finance at an appropriate tempo necessary safety measures.

Poor working conditions.

Failure to meet statutory requirements.

Inadequate design specifications or failures to meet or to sustain specifications for plants, materials and equipments.

Poor provisioning of expendable safety materials.

Production requirements being permitted to over-ride safety needs.

Ineffective security.

Failure to provide for standby emergency services, eg electrical generators.

Unscrupulous management or trade unions.

Certain important technical lessons have also come to light, as follows:-

Technical

Components crucial for safety must be designed so that malassembly during production or after maintenance and inspection is not possible.

Inspection sampling schemes for equipments, components and materials must take proper account of the hazardous effects of the consumers' risk; the *lowest* end of confidence limits must be acceptable for a particular application and give the necessary degree of reliability.

Siting of plants and processes must be satisfactory in relation to the *maximum credible accident*.

The philosophy for risk management must accord with the principle that, in spite of all precautions, accidents are inevitable. Hence the effects of a maximum credible accident at one location must be constrained to avoid escalating consequences at neighbouring locations.

Each location within a hazardous plant must be licensed by management for storage or processing only of specified quantities of hazardous materials. The licensed quantities must be prominently displayed. Subsequent relicensing for larger quantities must take place only when all the implications are known and regarded as acceptable.

No repairs or modifications to hazardous plants must be authorized unless all materials and methods employed comply with stated specifications.

All faults, accidents and significant incidents must be recorded and fed back without fail or delay to the Inspectorate.

Inspectorates must have delegated authority — without reference to higher management echelons — to shut down hazardous operations following any failure pending thorough evaluation.

Packaging and transportation criteria for hazardous materials must be satisfactory in relation to the maximum credible accident which could arise en route.

Where safety systems are duplicated (ie where the principle of redundancy is regarded as necessary) these must be completely independent and must not provide a common point or path where some event could put them both out of action simultaneously.

Designers of hazardous plants must seek to achieve systems which are so far as possible invariant under a maximum range of human errors, perverseness, indiscipline, inattention and stupidity irrespective of the excellence of any system of selection, education or training.

Appendix No. 5

ASBESTOS HAZARDS AND STANDARDS A STUDY IN THE ARBITRARY ACCEPTANCE OF HAZARDS.

by Anthony D. Woolf.

An object lesson in the failure of decision-making processes to incorporate adequate techniques or criteria for assessing the acceptability of risk, and the consequences of that failure, may be derived from the history of the asbestos processing and using industries in the UK.

Without attempting to survey the full history of control or to trace that history back to its origins, we can note initially the provisions of s.1(1)(d) and s.7 of the Factories and Workshops Act 1901. These required the employer to prevent the giving off or accumulation of any harmful dust or to render it harmless and to ensure that "in every room in any factory or workshop sufficient ventilation shall be maintained." Already the previous year the Home Secretary had appointed a special committee to enquire into such ventilation. In its second report published in 1907 the committee, consisting of J. S. Haldane, with an Inspector of Factories, the chief engineering advisor to the Inspectorate, and an Oxford pathologist, showed a grasp of techniques and needs which, in light of subsequent events, appears surprisingly advanced. After dealing with general ventilation, they turned to the question of dusts, gases and fumes, and wrote:-

> "As regards removal of dust, the standard of purity aimed at should always be sufficient to prevent injury to health and should also be such as to prevent inconvenience and enable those employed to be clean when they leave work, after washing, if necessary. Dust from the disintegration of hardstone, steel-grinding etc is extremely deleterious and the same may be said of dust containing any poisonous constituent, such as lead. In such cases the dust should, by special means apart from general ventilation, be entirely prevented from mixing with the general atmosphere of a room; and the same remark applies to all poisonous gases and fumes."

They added this illuminating footnote: "It is sometimes difficult to say whether the inhalation of a given variety of dust is definitely injurious. During our inquiry many experiments have been made by Professor Ritchie with a view to finding a means of experimentally distinguishing the more injurious from the less injurious dusts, but unfortunately no satisfactory results have as yet been reached." The committee accordingly treated all dust as dangerous unless and until the contrary was proved. Their report then detailed the highly developed theory and techniques of process isolation, control of air current direction and velocities and local exhaust ventilation and, noting the disadvantages and inadequacies of respirators as a means of defence, recommended a "no-dust" standard wherever possible. They demonstrated that, with careful design and maintenance of workrooms, processes, plant and systems of work, very high standards could generally have been achieved.

The 1907 report did not specifically mention asbestos, but, in 1906, Arribault had reported its lethality in France and this was brought out again by the UK Medical Inspector of Factories, Dr Collis, in the Chief Inspector's annual report for 1910. To some extent the lesson was apparently learned, for, in 1927, Dr Cooke, writing up a case of asbestosis in the British Medical Journal, said: "In up-to-date (asbestos) factories all machines are fitted with extractor covers and the dust removed." Whilst the truth of that statement is open to the gravest doubt, it shows what was then understood to be the requirement. In 1910 the asbestos weaving industry was in its infancy; by 1930 the International Labour Office was writing in its monumental encyclopaedia "Occupation and Health":-

"The weaving of asbestos has only developed importance during the last 20 years. Now that it is widely practised, the application of local exhaust systems during such processes as that above described is called for. All processes from extraction onwards unquestionably involve a considerable hazard, and American and Canadian life insurance companies generally refuse asbestos workers on account of the assumed deleterious conditions in the industry."

"The lack of more accurate and detailed data in medical literature regarding this industry in its various branches, including the utilization of by-products, is to be deplored in view of the self-evident importance of asbestos dust as a predisposing cause of pulmonary tuberculosis, more especially since the rapidly increasing development of industries utilizing asbestos adds greatly to the urgency of studying the conditions with a view to their amelioration".

"All provisions already described for withdrawal of dust should be enforced in this industry."

Thus a great hazard was by then universally recognised, the means for controlling it substantially in factories was available, widely publicised and required by law to be used, and the likelihood that the *use of asbestos products* involved a similar hazard was equally well appreciated. But, in that same year, the Home Office published the crucial report of Merewhether and Price on "The Effect of Asbestos Dust on the Lungs and Dust Suppression in the Asbestos Industry". This survey of 363 workers (out of an estimated 2,200 then employed in the industry) of whom 86.5% had been employed for less than 15 years, revealed a grave risk of fibrosis and associated chest diseases apparently related to length and intensity of exposure. The occurrence of cases after 12 and 15 months' exposure was recorded, and the need for further research was recognized, both into methods of control, and to determine the "harmful effects of comparatively low concentrations of asbestos dust." The report showed no grounds for departing from a "no-dust" standard, though it adumbrated the possibility that such grounds might eventually be established. But as the authors stressed, it was based on an extremely limited survey (confined to workers apparently exposed to no dust other than asbestos) and on extremely general and unsatisfactory data about their histories of work and exposure. The dustiest processes were identified, and the greatest incidence of disease in the sample was shown to have occurred in them, but that is about as far as quantification of the hazard could be taken in that report. Yet it appears that legislative and enforcement policies have been allowed to rest on that and two subsequent and equally unsatisfactory studies for nearly half a century since.

The Merewhether-Price report was published in March 1930; the Minister set up a Committee consisting of three representatives of the manufacturers and two members of the Factory Inspectorate whose recommendations, in their report made in April 1931, were published nearly verbatim as draft Regulations in July of that year. In the brief interim they had been "accepted" by a conference of manufacturers, and "discussed with representatives of the TUC", at whose instance three changes had been made. There was no public inquiry or wider consultation, and the regulations passed into law that year. There is no record of any attempt to enforce them anywhere in the country before the prosecution of Central Asbestos Co. Limited in 1964. (Brian Harvey, when he was Chief Inspector of Factories, brought one previous prosecution, but has published no details). Documents disclosed in various civil actions brought for personal injury damages since 1963 (when the limitation laws were reformed) and some published in the Parliamentary Commissioner's Report of 1976 show that Factory Inspectors used their powers of persuasion as best they could to bring about improvements in conditions, but flagrant infringements were widespread and everywhere tolerated. Powers of compulsion were were never used. The TUC's main addition to the regulations — compulsory use of impermeable sacks to contain asbestos — was almost universally ignored.

The original "no-dust" standard has been characterized as unenforceable and in itself a reason for non-enforcement by the Factory Inspectorate and the DTI in statements following the release of the Ombudsman's report. By 1960, however, the Inspectorate, with ministerial but not Parliamentary sanction, adopted the Threshold Limit Value (TLV) for asbestos published by the American Conference of Industrial Hygienists and accepted here by the Industrial Health Advisory Committee. The Ombudsman refers to this in his report as 'a "safe" standard of 177 particles of asbestos dust per millimetre of air' and it was doubtless assumed by many in the industry and the Inspectorate that this was indeed a "safe" standard. By contrast, study of the basis upon which it was set up reveals that it was correctly described in a 1967 Ministry of Labour booklet "Problems Arising from the Use of Asbestos" which stated (p.26): "Retention of this value in British practice is almost certainly unjustified, although in its favour originally, it did provide in the USA a guide level, *albeit an arbitrary one*" (my italics). And careful study of the introduction to the table of TLVs reveals that none of them have been scientifically established as *safe* levels, although the language is so opaque and convoluted as to give the impression that they have been so established to all but the most alert and suspicious reader. Equally, the Asbestos Information Committee (AIC) (the publicity organization of the industry) claims and implies throughout its published literature that safe levels of exposure are known. In any event, the adoption of the 177 particles standard was quite irrelevent to enforcement policy; no action was taken against Central Asbestos, for instance, when concentrations of 680 times that level were found, and the practical difficulty of monitoring was itself given as a reason for non-enforcement of the regulations.

Equally arbitrary was the adoption in 1969 of the current so-called 2-fibre standard on the recommendations of the British Occupational Hygiene Society. This standard was derived from the reported findings of two senior employees of Turner and Newall, one of the largest employers in the asbestos industry, on their own readings of X-rays of just 37 workers correlated with their own dust-sampling results. These were reported as showing only 3% with changes consistent with asbestos-induced fibrosis and it was assumed that a 2-fibre standard would reduce the risk effectively to nil. However, the standard does not pretend to take into account the risk of cancer, and it is understood that the underlying data (which has never been published), was not even independently checked. The X-ray interpretations are now said, on the admission of Dr Knox who provided them, to be wrong in 47% of the entire study, but the declared policy of the Factory Inspectorate is not to enforce the 1969 Asbestos Regulations unless that discredited 2-fibre standard is exceeded; and the AIC's publicity again suggests that no danger exists if the standard is met. There is no scientific basis for such an assertion.

Meanwhile, in the decades since the 1930s, the asbestos industry has grown into a giant, and has diversified its products and their application into every field of enterprise. Building design and materials supply, chemical and heat engineering and speed technology have developed to exploit and to depend upon asbestos products to the point that the entire urban population and, probably, the entire population uses asbestos products, and is exposed to the inhalation of asbestos fibres. Whilst a committee of inquiry into the hazards of using asbestos products was finally established in 1976, in response to widespread public concern, its independence has been seriously questioned and its freedom of action may be heavily circumscribed by the economic and political implications any adverse findings and recommendations would have. As distinct from the development of controls as a contemporaneous part of the development of a new technology, the task of bringing under new control a vast, deeply entrenched industry and the marketing and use of its products is an undertaking which may well be beyond the capacity of our constitution and machinery of government.

If that task is now to be embarked upon seriously in relation to asbestos, the most adverse assumptions of hazard must be made initially, for scientific data to justify less adverse assumptions does not exist and decisions cannot await the long programme of research required to produce them by reason of the long latent period of relevant diseases. Accordingly, a true exercise in judging the acceptability of risk is required in a situation and at a time when hazard is known to exist, the level at which it becomes negligible cannot be established, the widest review of social and industrial dependence on the hazard-producing substance is required, and the probable cost, including the cost of dealing with alternative hazards, would seem to require the entire resources available to government for its computation. The history outlined above is in marked contrast to the control of ammunition and explosives described in Appendix 4 to this Report, and that fact is itself a substantial reason for the current crisis of confidence. Public alarm about the record of past deaths in and around the industry, and the role that asbestos may have played and still be playing in the causation of cancer in the general population, shows that the hazard that has been imposed is not now acceptable. If it is to be said that nevertheless it must continue to be accepted at anything like the present level, the basis on which such a decision is to be taken must measure up at least to the rigours outlined in the main body of our Report.

Appendix 6.

A COMMUNITY RISKS ADVISORY SERVICE SOME PRELIMINARY PROPOSALS.

1. Introduction

It is not widely appreciated that the new legislation on Health and Safety is intended to cover a wider field than the workplace. Perhaps because of the urgency of improving industrial safety and also the uncertainties of creating a service for the far less structured hazards outside, there has not yet been any discussion of risk advice in the community. We came to this problem for its intrinsic importance, and also because the approach of community involvement and independent advisers seems particularly appropriate here.

So long as the inequalities of power among the different interests go unrecognized and unchecked, all control arrangements, however excellent formally, will tend to be rendered ineffective. Thus the record of "liaison committees" mixing industrial and local interests has not been significantly better than that of the Alkali inspectorate by which they are constituted. Unless representatives of the community at risk are independent of those creating the risk, and answerable to those they represent, they simply cannot do their job adequately.

Hence we recommend the establishment of a pilot scheme for a "community risks advisory service" which would serve to correct the imbalances we have discussed. The primary task of the adviser would be in the provision of information on all sorts of risks. This would range from relaying published statistics on well known hazards, to the sharing of skills in the work of monitoring or information retrieval. S/he would also give advice where appropriate, and when necessary, help to represent the community in joint discussions where risks are decided.

2. Relating to Existing Institutions

Because the risks experienced by members of the public are so diverse, the control of them is scattered and fragmented among a number of agencies, responsible to different sections of the government or of the Health Service. Each of them has a routine area of inspection and control, be it factories, shops, foodstuffs, traditional pollutants in water, infectious diseases and so on. Only exceptionally do they work by the principle of sensitivity to the emergence of quite new risks from environmental or technological developments, or have the means of detecting and eliminating them. Any community risks advisory service would need to cooperate with existing agencies, to use their good offices for information and technical services, and generally to be complementary to them in function. But this does not reduce the need for such an independent service.

3. Necessary Facilities

The main work of the adviser would be concerned with *information:* obtaining it from the relevant source and passing it on to those in need. Only occasionally would s/he be involved in direct negotiations on risks; the need for such methods arises only infrequently and even then the adviser should help the community to help itself, rather than becoming yet another expert acting on behalf of others. The tasks of providing information on risks are considerable; even the location of likely sources requires a considerable base of information. In the initial stages of this work much effort will need to be spent on the development of an information service, including location of sources and resolving the particular problems of secrecy, confidentiality and the cost of, or difficulty in, access. Some centralised register and exchange systems will need to be developed; but in all this work, the principle of low cost and informality should not be sacrificed to an abstract comprehensiveness or efficiency.

The problems of collecting data from monitoring operations are similar. The advisory service will need to have some equipment of its own, but should also be able to borrow more specialised equipment and to obtain laboratory tests and analyses from public institutions of one sort or another. The development of tests that are simple, yet reliable, adapted for use by many citizens rather than by a few experts, is an important part of the work.

Equipment and funds for collating and distributing information are essential for the involvement of the community in any programme of risk control. Even the office of the adviser should be planned with some care — preferably situated in the neighbourhoods most needing the service, open outside working hours, and with space for meetings and technical instruction.

4. Staffing and Finance

The initial development will require dedicated work by a few full-time people. But once established, the service should rely as much as possible on part-time workers, and on consultants engaged for particular tasks. Otherwise it will be more difficult for the service to avert the pressures constantly shaping it into yet another welfare bureaucracy.

A sample budget for the initial period of establishing a community risks advisory service for each of three years is as follows:

	£1,000's per annum
Adviser (full time; gross salary)	3.5
Secretarial staff (part-time)	2.5
Office rents & Overheads	3.0
Information & Reprographic work	1.0
Monitoring equipment	3.0
Analyses	0.5
Consultants fees	1.0
	£14.5

In later years, the costs of full-time staff could be shared among several centres, and the stock of equipment would only need repair and replacement.

In any place where a pilot scheme were launched, there would need to be some institution capable of offering assistance and advice in the early stages and also being able to make some guarantee for its continued existence should it prove successful.

Appendix 7.
THE USE OF MATHEMATICS IN THE ASSESSMENT OF RISK

by Dr. J. R. Ravetz.

The great mathematician Gauss once said that the "lack of mathematical culture is revealed nowhere so conspicuously as in meaningless precision in numerical computations". (1) Although this principle is appreciated by experienced scientists, it is hardly anywhere mentioned or amplified in books or courses for students or research workers. As a result, many scientists, along with the general public, develop quite erroneous ideas about the sort of information that is conveyed by a quantitative statement. It is not easy to point to an example where serious blunders in physical sciences or engineering resulted from abuse of numbers. This may be because such incompetence is swamped by more obvious sorts in each case that comes to light, or perhaps because it was never sought. But when we are dealing with statistical information, and making decisions on important policy questions, there is a strong possibility of serious errors resulting from misunderstood mathematics. For this reason I consider it worthwhile to review some of the problems of applying mathematics in this difficult field, discussing probabilities and modelling in some detail.

1. Quantitative data on rates of occurrence of undesirable events are at the foundation of any risk analysis. Unless we have some estimate of how frequently some component, system or monitoring arrangement has failed, we have little evidence for how likely it is to fail in the future. But the data themselves do not tell their story; they need interpretation in terms of a statistical argument. For example, suppose that on a particular test a component fails 1% of the time. Are we justified in concluding that it is 99% safe for the future? Perhaps, but *how* justified? One failure in a hundred trials does not seem to tell us as much as 10 failures in a thousand. Indeed, it does not; if the consequences of the failure would be serious, we will require more confidence in the 1% estimate than is provided by the single failure in the small sample. When policy decision are based on calculated probabilities, the bare numbers should be supplemented with "confidence limits". These describe the range over which the given number is the "reasonable" one to us; they are derived

by a mathematical theory, and depend on the size and shape of the sets of data, as well as on unverifiable assumptions about the "universe" of possible data from which they are drawn.

An example of very practical relevance is furnished by Lowrance, commenting on an FDA document on carcinogens:

> "Even with as many as 1,000 test animals, and using only 90% confidence limits, the upper limit revealed by negative experiment (one revealing no tumours) is 2.3 cancers per 1,000 test animals. No one would wish to introduce an agent into a human population for which no more could be said than that it would probably produce no more than two tumours per 1,000. To reduce the upper limit of risk to two tumours per one million (at confidence limits of 99.9%) would require a negative result on somewhat more than three million tests animals." (2)

Every competent statistician knows that each of his tests involves the possibility of two sorts of error, giving conclusions that are too "optimistic" or too "pessimistic". He adjusts his tests to give a result with a proper balance between the likelihoods of the two errors, his choice depending on the policy context of the test. What he cannot do is to eliminate both sorts of error. So every statement of a probability about events is itself qualified by the likelihood of its being correct. This may be disappointing for people who expect simple certainties in numbers; but it is inevitable and should be better known.

When probabilities of hypothetical events are estimated, and then compared, the range of magnitudes of the resultant probability of a realized hazard may be uncomfortably large. For example, the "Rasmussen" report (3) considered the possibility that in a core meltdown of a nuclear reactor, the fuel might fragment to such a degree as to cause a "steam explosion" on contact with the cooling water. The probability of this was estimated at 1/10; and the probability of the resulting explosion fracturing the protective vessel was similarly 1/10. Thus this particular consequence had only a 1% chance, contingent on the already unlikely core meltdown accident; and so it was in the "negligible risk" class. But, as the report admitted, there is very little experimental evidence on the pheonomenon, either the conditions of its occurrence or its explosive force. It appears that 1/10 was used as a standard transition probability. Thus a two-step process is reduced by 100; but had a transition probability of 1/3 been used, the reduction would have been 1/9, and the risk might not have been so easily considered negligible for policy purposes.

An awareness of the imprecision of all estimates of probabilities is quite essential when these are fed into complex mathematical computations. "Errors", or more properly, "inexactness" in quantitative statements can be increased enormously by some operations, so much so that a result of a computation may be actually devoid of meaning. The most dramatic simple example is the computation of the algebraic expression 1/(b-a); if these differ by, say, 5%, and are each inexact by 1%, then the result of the computation may vary by more than a factor of 2. (For example, if a and b are 95 and 100 respectively, then 1/(b-a) can vary between 1/(101-94) and 1/(99-96), or 1/7 and 1/3; or .14 and .33 respectively. If a and b differ by 2% or less, then the answer is totally indeterminate, since (b-a) can be zero or a negative, and 1/(b-a) infinite or negative.) Therefore to the extent that mathematical models involve anything but the most straightforward addition, multiplication or comparison of probabilities, the confidence limits of these quantities must be seen to be as strict as necessary. In some fields of safety engineering, this policy is adopted as a matter of course; in others it is possible to explore extensively in sophisticated mathematical discussions of hazards, risk, imputed values of life, and related topics, never seeing a sign of awareness of the need for confidence limits on probabilities. In the latter case, the expertise that is displayed in all the calculations is not genuine.

Another sort of important "spread" of numbers describing hazards, is the severity of risks to people in "the tail of the distribution". There may be situations where most of the population is "safe", some 10% are at risk, and 1% are severely exposed. A bare statement of average effects can then by implication conceal the danger; some measure of the variation of the risk should always be included.

Many hazards are described by extremely small probabilities, and these need to be handled with care lest meaningless calculations and conclusions result. Since the probability of occurrence of two independent events is the product of their separate probabilities, it is easy to produce statements of risk of say 10^{-5} events per year. But a probability is meaningful only if the event in question could reasonably be expected to occur; and one per 10^{15} years is not such. Such astronomically large numbers are obtained by multiplying the low but meaningful probabilities (say 10^{-5}) of several events which are jointly necessary for the realisation of a particular hazard along a particular pathway. The fallacy lies in neglecting other parallel pathways, which may have been omitted from the model because of being improbable in comparison to any single one of these events. Thus if one is to discuss compounded accidents to installations, calculated at 10^{-10} events per year, then one should include earthquakes and airplane crashes, even though their probability may be only 10^{-7}.

As a practical guide, risk analysts generally accept 10^{-5} to 10^{-7} events per year as the very lowest limit of realistic assessment; less than that is simply "negligible", and is not to be calculated with. But this "lower limit of computability" depends on the hazard situation; it is clearly lower for the regularly repeated hazards as aircraft, car or personal accidents, than for hypothetical industrial mishaps. In the latter case, 10^{-6} is sometimes taken as the lower limit of computability. But this is only a factor of 10 less than a risk level that is commonly taken as a design target. Can one be sure that a probability computed at 10^{-5} events per year is always meaningful? It may be, but only if the person doing the calculation is skilled and responsible. Thus here as elsewhere, the right use of quantities is a skill involving qualitative judgments.

2. Mathematical models have come into wide use for the description of the complex causal networks of major industrial hazards. They have provided a basis for disciplined and realistic thinking about such hazards, making an advance beyond guesswork and uncontrolled assumptions in design policies. In them, the different possible initiating events and their connections are indicated, and probabilities of occurrence are assigned. In spite of the abstract, logical appearance of such "hazard trees" (to use the name of a common variety), they depend very strongly on the intuition and skill of the modeller, for their value as an analytic tool. Debate over the quality of any particular model will then range over issues where quantitative experiments cannot decide the issue, but only experience and judgment.

The modeller must reproduce both technical and probabilistic features of the hazard as faithfully as is required for the policy conclusions to be reached on the basis of his analysis. For the structure, he must decide how far back to go into remote causes and unlikely connections. To include everything that logic permits, would produce an unwieldy and confusing model. In particular, since it is very difficult to assign a probability of failure of monitoring (4), this cause tends to be omitted. But each act of omission of a conceivable cause entails a policy consequence: "this is not to be worried about". If the context changes as in the case of terrorism or sabotage of nuclear installations, excluded from the Rasmussen report but now recognized by the British authorities, then the analyst must revise his calculations if he is to remain credible.

The assignment of probabilities of failures is equally demanding. Where there is much data on standard components, that are to be used in nearly unchanged conditions, then confidence-limits can be close, and causal networks established with considerable precision. An example of this is the case of aircraft take-off and

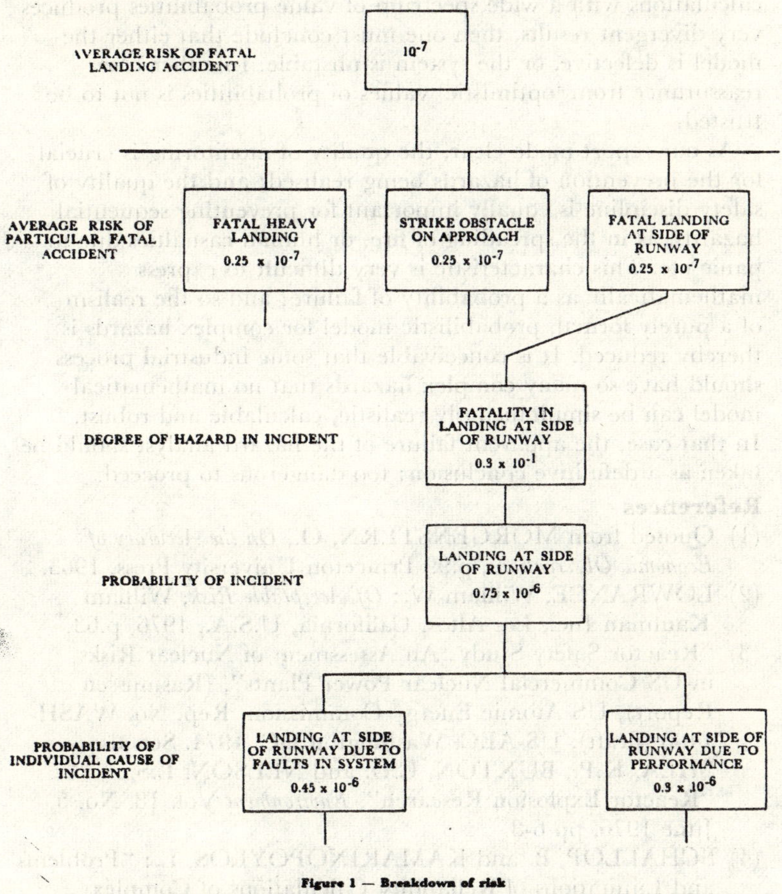

Figure 1 — Breakdown of risk

From: D. V Warren — Safety assessment of systems for landing aeroplanes in bad visibility. (In: Systems reliability in high risk situations — One day seminar on Thur 10 Oct. 1974) UKAEA, Risley.

landings; Figure 1 shows how the risk can be analysed into simple and regular elements.

But to the degree that the parts and the uses are new, the probabilities should lie in a broader interval of confidence-limits. Paradoxically, such may be the case where precision is most needed, in the calculation of hazards in complex, high-energy systems. With these less precise probabilities, the model is tested in a new way: whether it is sufficiently "robust" against changes in numerical values. This depends on its underlying logic, which in turn represents the structure of the hazards. If the running of calculations with a wide spectrum of value probabilities produces very divergent results, then one must conclude that either the model is defective, or the system is unstable. In either case, reassurance from 'optimistic' values of probabilities is not to be trusted.

As our report made clear, the quality of monitoring is crucial for the prevention of hazards being realised; and the quality of safety discipline is equally important for preventing sequential hazards, as in the spreading of fire, or human casualties due to panic etc. This characteristic is very difficult to express mathematically as a probability of failure; and so the realism of a purely formal, probabilistic model for complex hazards is thereby reduced. It is conceivable that some industrial process should have so many complex hazards that no mathematical model can be simultaneously realistic, calculable and robust. In that case, the apparent failure of the hazard analyst should be taken as a definitive conclusion: too dangerous to proceed.

References

(1) Quoted from MORGENSTERN, O., *On the Accuracy of Economic Observations*, p.99 Princeton University Press, 1963.

(2) LOWRANCE, William W.: *Of Acceptable Risk*, William Kaufman Inc., Los Altos, California, U.S.A., 1976, p.63.

(3) "Reactor Safety Study: An Assessment of Nuclear Risks in US Commercial Nuclear Power Plants", (Rasmussen Report), US Atomic Energy Commission, Rep. No. WASH-1400 (draft), US-AEC, Washington DC, 1974. See also SHEA, K.P., BUXTON, C.D. and NELSON, L.S.: "Reactor Explosion Research", *Environment* Vol. 18, No. 5, June 1976, pp 6-8.

(4) SCHALLOP, B. and KAMARINOPOYLOS, L.: "Problems and Limitations of Reliability Calculations of Complex Systems", in *Principles and Standards of Reactor Safety*, IAEA-SM-169, Vienna, 1973.